Fiery Trial

FIERY TRIAL

Judge James H. Lincoln (ret.)
James L. Donahue

Historical Society of Michigan
Ann Arbor, Michigan

Contents

List of Illustrations

Publisher's Note

Fiery Trial is the first volume published by the John W. Gillette Publishing Fund of the Historical Society of Michigan. Established in 1983 by the Society's Trustees, the Gillette Fund honors the memory and work of John W. Gillette of Berrien Springs who served as a trustee of the Society from 1970 to 1976, and as its president from 1974 to 1976. Before completing his tenure on the Board, he established the Michigan Heritage Library, a joint venture between his Hardscrabble Books, Inc. and the Historical Society of Michigan. The Heritage Library features reprints of notable books from Michigan's history. Mr. Gillette was working on the production of the third volume in that series at the time of his death in August 1983.

The Society's Gillette Fund will publish original manuscripts on various topics in Michigan's history. It is a revolving fund, so proceeds from the sale of one volume will be used to support the publication of the next one.

The Trustees of the Historical Society of Michigan gratefully acknowledge the support of the following foundations which generously contributed to the establishment of the Gillette Fund and the publication of *Fiery Trial*:

- The Knight Foundation, Akron, Ohio
- The Earhart Foundation, Ann Arbor, Michigan
- The William Jr. and Dorothy Zehnder Foundation, Frankenmuth, Michigan

Thomas L. Jones
Executive Director
Historical Society of Michigan

Introduction

Researching and writing a book about the terrible fire that swept Michigan's Thumb Area in September 1881 made this 100-year-old catastrophe seem too real.

Jim Lincoln and I discovered that we both experienced similar visions of great forests engulfed in flame, people and animals fleeing in wagons and on foot to Lake Huron or some of the local streams and great balls of burning pitch falling from an ominous black sky as the fire approached.

It was difficult to drive through the countryside without thinking about how severe the fire was at this point, or the people who died at that place. We were haunted by the ghosts of the past.

Now that we have finished our work, we both feel that somehow we have personally lived through that fire. The fact is that we were writing a very personal book about a terrible event that happened to the parents and grandparents of friends and neighbors in an area where we both were born and raised. The mind pictures were made even more vivid when we read personal accounts from letters and newspaper articles uncovered during Jim Lincoln's three years of careful research. These accounts are included in this book.

For our readers' benefit, perhaps the area known as the Thumb, and the events that occurred there, need to be put into perspective.

The lower portion of Michigan is shaped like a jagged mitten, surrounded by Lake Michigan on the west and Lake Huron on the east. Lake Huron wraps itself around an obtrusion of rich, flat lowland. It is not quite a peninsula, but on maps it looks like the thumb of the mitten. This area, which includes Huron, Tuscola and Sanilac Counties, is commonly known to Michiganians as the Thumb.

The great fire that occurred here in 1881 was the second of two terrible forest fires within a decade.

The first fire happened on October 8, 1871 and was part of a great holocaust that claimed thousands of acres of forestland in Michigan and Wisconsin, plus the City of Chicago. However, its destructiveness was less than the 1881 fire which claimed more lives and property. That first fire brought an end to local lumbering operations and led to the sale of the land in small parcels. Within 10 years the woods were filled with settlers and small crossroads communities.

The 1881 fire, fanned by 40-mile-an-hour winds from the southwest, was a killer. As the fire grew in intensity, it spawned even stronger winds and possibly even tornadoes, as indicated in some of the accounts we uncovered.

Nearly 300 people died and many others were seriously burned. Few trees and buildings were left standing. More than 2,000 square miles of forest and towns became charred wasteland within five hours. The winds were so powerful they toppled trees, tore roofs from buildings and tossed people in the air like pieces of paper.

The scars from that terrible day are still found: old cedar tree stumps, charred logs used for timbers in homes built after the fire, and most especially in the minds of the children and grandchildren of the survivors. They still tell the stories.

James L. Donahue
April 28, 1984

The Great Fire of 1881 has been a story waiting to be told. It is true that much has been written about this holocaust that swept over nearly two thousand square miles of the Thumb of Michigan a century ago. However, prior to the assembly of the material for this book, no one had ever made such a comprehensive collection of the hundreds of individual personal experiences of the people who were here in the Thumb on September 5, 1881.

Time has been running out. There are still hundreds of people in the Thumb area living on the same farms or in the towns where either parents or grandparents lived at the time of the fire. These people have heard firsthand accounts from their parents or grandparents of what happened on their farms during the fire.

Some of the people who contributed to this book have passed on before it could be published. Much of the information in this book simply would not have been available even by the year 1990.

My interest in the Great Fire is rooted in my early boyhood in the 1920s. My grandmother Philura Lincoln (1852–1939) lived on our family farm from May 1881 to her death in 1939. When I was a young boy, she repeatedly told me her eyewitness account of lighting lamps and lanterns in the intense darkness of midday and of the heavy coating of white ashes that covered the ground a few hours before the fire itself. I recall most vividly her telling about a large grove of maple trees on our farm next to Lake Huron. The Indians used to come yearly to this spot for their annual harvest of maple syrup. The grove of maples was destroyed in the fire and the Indians came no more.

At the time she related these events to me, our home was located where the grove of maple trees once stood. Although over forty years had passed since the Great Fire, she spoke of it as though it had happened only yesterday. The Great Fire seemed very near to me in time and place. I would imagine what it would be like to be on that spot on September 5, 1881. To remain there would have been to die. All along Lake Huron shores hundreds of people and thousands of animals sought shelter in the cool waters of Lake Huron. The solution for those who lived next to the shore was simple. But those to the west of our farm in Paris, Rubicon, and the west end of Sand Beach Township died in large numbers.

Our farm home still had kerosene lamps and lanterns similar to those used in 1881. Electricity did not come to our home until after I left for The University of Michigan in 1934. The stone well by our barn was identical to the stone wells that gave shelter and life to hundreds of settlers during the holocaust. The plumbing was still behind the lilac bush. The farms of the 1920s and the 1930s were similar to the farms that were here in 1881. We still used horses to pull our plows and do our work. A thousand years of change seemingly took place from the time I left the farm in 1934 until I returned in 1977. It is important to understand that most of the people who contributed to this book identify very strongly with what was here in 1881. Many of us have lived in both the old world and the new.

Over three years were spent collecting material from libraries, the Red Cross, the National Archives and from newspapers. Many people were interviewed and cemeteries and sites of burned out villages were inspected. Approximately 500 letters were sent to centennial farm owners in the Thumb area requesting information about the Great Fire. Newspaper articles together with letters brought an almost unbelievable response and a rich harvest of accounts of individual experiences.

All resource materials and documents collected to write this book will remain with the Historical Society of Michigan.

In 1979, Jim Donahue, a history buff and a reporter for the *Port Huron Times Herald* wrote an article about my project of collecting information about the Great Fire. This one article resulted in several dozen letters and even a greater number of phone calls from people in the Thumb area who wanted to give me information concerning the fire. Jim has written a number of newspaper articles on the Great Fire. In the spring of 1981, I turned enough resource material over to Jim Donahue to write ten books! Jim was given the difficult task of condensing the contents of approximately 90 letters and other lengthy material into a usable length for the book without distorting their meaning or losing their flavor. After 25 years in elected office with daily contact with news reporters, I can state that I never met a more accurate news reporter than Jim Donahue. His style of writing and approach is that of an investigative reporter. Some accounts of individual events were eliminated from this book because they were not credible or were not properly substantiated.

There was often evidence or materials supporting more than one version of some event that occurred during the fire. We resolved them by clearly indicating in the book whenever there is conflicting evidence concerning the event.

There are at least a hundred books in my library concerning the Civil War. There are often widely conflicting accounts of the same event given by two or more reputable authors or historians. However, these reputable historians often fail to indicate that there is stong conflicting evidence about a particular event. *Fiery Trial* clearly indicates that there are conflicting accounts of the same event.

This book is not historical fiction. *Fiery Trial* is well documented history. We are proud that the Historical Society of Michigan has sponsored *Fiery Trial*.

This project was not undertaken solely to produce a book. It was a pilgrimage into the past and I have tried to personally relive what many of the people in the Thumb area experienced during the Great Fire. I would walk down some wooded lane or find myself in some place of special significance through which the fire had swept, and imagine that I was in that place on September 5 1881. I have climbed down into the old stone wells, helped to carry water to the roofs of burning buildings and houses, fled down country lanes and heard the cries of burning animals.

Fiery Trial is an invitation for you to take a pilgrimage into history and relive the experiences of the settlers who were here on September 5, 1881.

The survivors buried their dead, nursed the burned and injured back to health, and rebuilt their homes. A better civilization rose from the ashes and ruins of the Great Fire.

James Lincoln
April 28, 1984

Acknowledgments

The writing of this book could not have been accomplished without the assistance of many people who took the time to answer public appeals for stories and records pertaining to the 1881 fire.

The following list cannot possibly include everyone who helped. Some information came to us by telephone and other stories were passed in coffee shops.

We also know that some stones have been left unturned. There are still more stories to be told. Our purpose was to collect and record some of the stories before they were lost forever.

Our thanks go to the following people:

Donald R. Tippie, Executive Director of the St. Clair County Chapter of the American Red Cross, Port Huron, Michigan, who was our chief source of reference material from Red Cross sources. He furnished material from the Port Huron Chapter and contacted Robert A. Howard, Director of the Office of Public Affairs, American Red Cross, Washington, D.C.

Martha M. Bigelow, Director of the Michigan History Division, Michigan Department of State, Lansing, Michigan, who furnished resource material and photographs; librarians Joan Dickinson and Mary Willett of the Harbor Beach Library, Harbor Beach, Michigan who gave us much assistance and compiled a considerable amount of reference material relating to the Fire of 1881; Margaret Ward of the

Burton Historical Collection, Detroit Public Library, Detroit, Michigan; Richard Doolen of the Bentley Library, Michigan Historical Collections, The University of Michigan, Ann Arbor, Michigan; the Bad Axe Public Library, Bad Axe, Michigan; and the *Port Huron Times Herald*, a newspaper that has written more about the Fire of 1881 than any other newspaper. These newspaper accounts appeared not only immediately after the fire, but over the past century as well. One hundred years later the topic is still newsworthy.

Letters and special assistance: Resource material was received in response to articles appearing in newspapers throughout the Thumb area about the authors and their research. One very productive source of material came from a letter sent by Judge Lincoln to over 200 Centennial Farm owners in the Thumb area requesting information about what occurred on these farms during the Fire of 1881.

We wish to thank:

Anson Babcock, Marysville, Michigan
Mrs. Dan Badgley, Port Austin, Michigan
Wallace Balhoff, Sandusky, Michigan
Mary Lou Baron, Grand Blanc, Michigan
John Bell, Marine City, Michigan
Ruth Benedict, Brown City, Michigan
Arlene (Root) Bourke, North Branch, Michigan
Beatrice Brandow, Harbor Beach, Michigan
Dorothy Brandt, Midland, Michigan
Dorothy Brind, Gagetown, Michigan
Robert Brobst, Harbor Beach, Michigan
Edgar W. Brown, Ann Arbor, Michigan
Linda Brown, Sandusky, Michigan
Myrtle Sparling Brunk, Tyre, Michigan
Eugene G. Burgess, Capac, Michigan

Dale Burley, Harbor Beach, Michigan
Mrs. BlenFord Campbell, Fairgrove, Michigan
Garnet Chase, Port Huron, Michigan
Rudolph A. Clemen, Washington, D.C.
Fandira Connelly, Harbor Beach, Michigan
Ida Olive Cowley, Lexington, Michigan
James E. Eddy, Englewood, Florida
Mr. & Mrs Herbert M. Elder, Deckerville, Michigan
Hildred Endershe, Owendale, Michigan
Mrs. Harold Evans, Clawson, Michigan
Mrs. Robert D. Falk, Port Huron, Michigan
Margaret Farley, Almont, Michigan
Ruth Foe, Harbor Beach, Michigan
Arlee Freiburger, Snover, Michigan
W. Fricke, North Branch, Michigan
Frances Gainor, Harbor Beach, Michigan
Bruce Gee, Cass City, Michigan
Stanley S. Graubner, Caro, Michigan
Mrs. Carl Harlet, Flint, Michigan
Mildred Harneck, Marlette, Michigan
Arlan E. Hartwick, Cass City, Michigan
Lorna Heath, Port Austin, Michigan
Lura Hedley, Caseville, Michigan
Theo C. Hendrick, Cass City, Michigan
William Hirons, Brown City, Michigan
Mrs. John Hopkins, Harbor Beach, Michigan
Mrs. Maurice Howell, St. Helen, Michigan
Matilda Irwin, Downington, Michigan
Allen Jirasek, Harbor Beach, Michigan
Monica Jirasek, Harbor Beach, Michigan
Dan Kanaby, Port Hope, Michigan
Charles F. Kempf, Port Huron, Michigan
Bernice A. Kennard, Vassar, Michigan

William Kitchen, Cass City, Michigan
James Kroetsch, Saginaw, Michigan
Harvey H. Krohn, Port Huron, Michigan
John E. Krug, Bad Axe, Michigan
Carl and Linda Krumenacker, Ubly, Michigan
Mary Lawler, Carsonville, Michigan
Ethel Lawrence, Port Huron, Michigan
Louise Lapp, Deckerville, Michigan
Geore P. Lawson, Port Sanilac, Michigan
Willis Leipprandt, Pigeon, Michigan
Kim Lincoln, Harbor Beach, Michigan
Lawrence J. Lincoln, Alexandria, Virginia
Barbara Loree, Port Huron, Michigan
Mrs. Ralph Lyons, Silverwood, Michigan
Ruby Maschke, Port Hope, Michigan
Elsie Matthews, Port Hope, Michigan
Olive E. McAllister, Port Austin, Michigan
Janet McGuire, Gagetown, Michigan
Maude Merithew, Raleigh, North Carolina
Joseph A. Messing, Minden City, Michigan
Clayton Milis, Marlette, Michigan
Samuel A. Minard, Zanesville, Ohio
Carmon L. Moore, Sandusky, Michigan
George E. Murray, Lapeer, Michigan
Robert Murray, Applegate, Michigan
June Nelson, Caseville, Michigan
Mrs. Frank Nolan, Port Sanilac, Michigan
Doris M. Norris, Yale, Michigan
Martin Olshove, Lexington, Michigan
Rosie Pattison, Port Austin, Michigan
Florence Peters, Lincoln Park, Michigan
Mrs. Willard Phelps, Mayville, Michigan
William Poison, Lapeer, Michigan

William A. Poldi, Peck, Michigan
Norma Pringle, Sandusky, Michigan
Mrs. James Pyle, Goodrich, Michigan
Beulah Eastman Racely, Port Huron, Michigan
Mrs. L. H. Riseborough, Port Huron, Michigan
Ellen Ryan, Capac, Michigan
Earl Schave, Port Hope, Michigan
Anna Schmidt, Sandusky, Michigan
Irene and Leland Schmucker, Harbor Beach, Michigan
Ted and Ella Schubel, Minden City, Michigan
Vernieta L. Scott, Jackson, Michigan
Ella Shanks, Carsonville, Michigan
Paul Soini, Bad Axe, Michigan
Mrs. Robert Snow, Imlay City, Michigan
Lynn Spencer, Ubly, Michigan
Mrs. Leonard Stern, Capac, Michigan
Leona Stillwell, Levering, Michigan
Charles and Helen Swayze, North Branch, Michigan
Stella B. Teets, Peck, Michigan
Caroline Thayer, Palms, Michigan
Naomi Therrien, St. Ignace, Michigan
Nettie VanRaaphorst, San Jose, California
Warren VanHorn, Vassar, Michigan
Mrs. Florence Vincent, Yale, Michigan
Glen Wakefield, Kinde, Michigan
Mrs. Claude Wood, Fullerton, California
Elizabeth Yeo, Marysville, Michigan

Donald M. D. Thurber, Grosse Pointe, Michigan merits special acknowledgment. He first suggested that the Historical Society of Michigan sponsor and publish the *Fiery Trial*. He worked out all arrangements to bring the project to a successful conclusion.

The Fire of 1881
—anonymous

'Twas on a summer morning,
   When everything was dry,
The tempest gave them warning
   That fiery darts would fly.

Come out with all your forces,
   Cried out the aged sire,
Prepare and take your places
   To fight the raging fire.

We'll fight and fight with reason
   Save everything we can,
For times are hard this season
   With us in Michigan.

Prepare yourselves with water,
   Have plenty at command
For great will be the slaughter
   This day on every hand.

Hark! Hark! The wind is raging,
  It tells a dreadful tale,
Ten thousand fires are blazing
  All round us in the dale.

Again the wind came driving,
  Time after time the same,
And kept those fires thriving
  'Til all was in a flame.

Fire like distant thunder
  Was heard on every hand,
A–burning things to sunder
  All o'er the timbered land.

And through the fields a–creeping
  A–flying with the gale,
The grain and buildings sweeping
  All o'er the hills and dale.

Loud cries in every quarter,
  In every dale and glade;
Some cried aloud for water
  And some would cry for aid.

Those fiery darts have found us. . . .
  O, God, what will we do?
And when they close around us
  We never can get through.

They tried those fires to master
  But could not stand the heat;
They still came thick and faster
  And forced them to retreat.

They planned in great disaster
  To shun the fiery grave;
All around the gloomy border
  Their precious lives to save.

But some so crazy–minded
  They rambled to and fro
And by the smoke so blinded
  They knew not where to go.

In holes where they were driven
  Would they sit down and cry,
And make their peace with heaven
  And then lie down and die.

The tempest loudly thundered;
  The fire refused to yield,
Until about three hundred
  Lie died upon the field.

With many sheep and cattle,
  And beasts of every kind,
Fell in that fiery battle
  And all to dust consumed.

The mailman fixed and started,
  Marlette he bid adieu,
Not thinking when they parted
  Of flames to travel through.

And now in that burnt region
  Resigned his fleeting breath
And now in that burnt region
  Poor Humphrey sleeps in death.

Those flames kept on their course
　Until the close of day;
The wind withdrew its forces
　And gently died away.

The smoke in fiery masses
　All settled to the ground,
And filled the vale and passes
　For many miles around.

The place was dark and dreary
　With smoke and fire combined,
The people sad and weary,
　Half smothered, starved and blind.

And in such wild emotion
　With many sighs and tears,
Cried out in sad devotion
　And praised their God with cheers.

Their souls were filled with gladness
　And many cares had fled
But clouds of grief and sadness
　Still hung about their heads.

Their fruits of all their labor
　In places on the shore,
With many friends and neighbors,
　All burned to be no more.

And they were left to wonder
　With loads of care and grief,
And o'er their losses ponder
　Till they could find relief.

And many in that number
　In little groups around,
All took their evening slumber
　In places on the ground.

The smoke and fire were blended
　And dreadful was the flow;
The day began and ended
　With horror, care and woe.

Long, long will they remember
　The smoke and fiery hum,
The fifth day of September,
　The year of eighty–one.

Beneath the Wild Cherry Tree

The race is not to the swift,
Nor the battle to the strong. . . .
But time and chance happenth to them all.
Ecclesiastes 9–11

The Hillside Cemetery is located three miles east and one mile north of Argyle. Directly north of the cemetery is an undeveloped state game area. Roads that lead through this state land are unimproved trails, turning and twisting through the trees, and they must look similar to the oxtrails that were called roads here a century ago.

I first visited the place in early September 1979. The blackberry season had peaked. The trees and vegetation were losing their lusty summer green but nature had not yet acquired the bright colors of autumn. This is a gently muted time of year when nature seems to pause between summer and fall. Time has healed the landscape and restored the woods that perished in the fire that swept this peaceful spot a century ago.

The marker of John and Susan Seder Cole and children commemorates these important and compassionate Michigan pioneers and the work they did in the aftermath of the Great Fire of 1881. Courtesy of Ed Lincoln.

It was the best of seasons to take a stroll along the trails through the woods and some of the open glades.

I wandered from the trail in a pleasant open glade, and there sat down on the grass to soak up the sunshine and muse about what happened in this place so long ago. It was not difficult to imagine that I was projected back through the century of time, for this woods and its pleasant scenes would have been much the same.

It is now a hot September day in 1881. I watch as the sun dissolves and becomes hidden in a huge cloud of black smoke that is traveling eastward high above the earth, and miles ahead of a great fire. A deep, dark twilight settles ominously over the woods and glade and midday. The wind is suddenly no longer gentle. It soon rises to gale force and the quiet woods are filled with the wind–tossed branches of trees and the roaring of rushing winds. The air becomes filled with white ash particles and some burning embers that have traveled on the wind ahead of the fire and they now begin to fall on the grass near at hand.

There comes a rush and scurrying of wild life fleeing eastward from the woods past me across the open space. There are dozens of rabbits, raccoons, snakes and an equal number of deer sweeping across the glade in great leaping bounds. A large black bear followed by two cubs cannonball across the clearing.

As they flee, they pay no heed to me, nor to each other. This is truly Judgment Day. For in this time and place the holocaust will sweep most of the landscape clean for miles in every direction: south 30 miles into Lapeer County; west 20 miles to Gagetown; north over 30 miles to the tip of the Thumb and east 15 miles to Lake Huron.

The cool waters of Lake Huron will save the lives of many hundreds of people and tens of thousands of wild and domestic animals. Lake Huron is too far away to be of help in *this* time and place. The only hope of survival now is to get to some open field and hope to somehow suffer through the intense heat and choking smoke and flying embers.

In this final hour, when humans and animals face their last extremity, there will be a drawing together. A dead woman will be found in an open field with her two dead children drawn close to her. Close by will be a large dead black bear that preferred the company of human companionship than to perishing alone.

The wild animals are accompanied in their headlong flight by thousands of birds that fill the air. Their shrill cries are mingled with the roaring winds. There are crows, hawks, bluejays, starlings, doves, quail, robins, bluebirds and dozens of other varieties of songbirds. They will die by the thousands in the smoke and flames and thousands more will fly out over Lake Huron and, after flying for hours without finding a place to rest, will finally drown in the lake.

A hundred species of flying, leaping, crawling and burrowing things speed past me in a race with a fiery death. Then suddenly they are gone and I am alone except for a huge owl that silently sweeps out of the west woods and glides swiftly close over my head.

It takes all my strength to stand against the wind and look westward across the clearing. There is the gloom of heavy twilight. There are strong gusts of wind, ashes and fiery embers hurtling eastward. There is a sound of branches breaking and crashing to the earth and the roaring wind seems to engulf me. A bright glow now appears through the haze and gloom above the western treetops.

I watch as a team of galloping oxen emerges from the woodland trail dragging an oxcart. A young man is standing in the cart shouting and trying to lash the oxen to greater speed with a long black whip. Foam is spewing from the animals' mouths and nostrils. Clearly they have come a long way and their race is nearly run.

The cart is overcrowded with five people along with the young man who it trying so desperately to get more speed out of the spent oxen. There is a woman of middle age holding a boy of perhaps two years of age in her arms. Another woman in her twenties clutches a young boy to her breast. There is a young girl of perhaps eight years of age.

I drop to my knees in the face of the gale force winds and watch the oxcart pass within 20 feet along the trail. They turn their set and terror stricken faces toward me as they pass but there is no sign that they have seen me, except the young girl who turns to the rear of the cart after it has passed. She holds to the side of the cart with one hand and waves to me.

They move east across the glade into the gloom of the woods and disappear into eternity.

To remain in this small opening in the woods is certain death. I have remained too long. It is now too late to flee a holocaust that can overtake a galloping horse.

Is it better to die here or run to the woods to the east and die there? One can fight and die on a thousand battlefields. One is fortunate if he can chose the time and place. This time and place is good enough for me.

I watch as the haze and gloom is dispelled in an onrush of blinding light. There is a mighty roaring as the fire crashed through the woods from the west to the edge of the glade. Flames leap 100 feet skyward. The blasts of wind carry a

One of the few landmarks that remain of the Great Fire of 1881 is the Hillside Cemetery located three miles east of Argyle. After the fire, John M. Cole buried six of his neighbors on his farm. He also planted a small cherry tree switch on the grave. The tree is over a century old today. Courtesy of Ed Lincoln.

The Stone marker beneath the wild cherry tree in the Hillside Cemetery marks the graves of six people who died in the Great Fire of 1881. Courtesy of Ed Lincoln.

searing, white-hot heat. The grass about my knees explodes in flame.

> I know it is over, over
> I know it is over at last
> The voices of freemen and lovers
> The sweet and bitter have passed.

I travel back down the time tunnel to this century and stroll back along the trail across the open glade and through the woods and pass through the front gate of a cemetery.

I am attracted to the west end of the cemetery by a huge wild cherry tree. Sheltered by its branches is a grave and a plaque. It reads:

"In memory of pioneer settlers, victims of the forest fire of September 5, 1881.

Emma Palmer, age 52	Neil Erhart, age 30
Mary Ann Erhart, age 24	Clare Howard, age 8
William Erhart, age 2	Clare Erhart, age 2 days"

The cherry tree in Hillside Cemetery is about as old as the cemetery. Both were started by John M. Cole, a pioneer settler and civil war veteran who lived in Argyle Township, fathered 14 children, and with his remarkable family, survived the great forest fires of 1871 and 1881. His wife was Susan Seder Cole.

When John Cole was plowing around the grave site he nearly uprooted a small cherry switch. Maybe he had a sense of history. For some reason he saved this little twig and saw that it was planted firmly.

After the fire, Cole collected the bodies of six of his neighbors, made pine coffins and buried them in a grave less than 100 yards west of where the Cole monument now stands in the cemetery. He donated the land for use as a cemetery and it continues to be maintained by the township to this day.

The great tree, now 100 years old, shades the gravesites of those burn victims.

The six bodies Cole buried were of people from one family who died at a farm in Wheatland Township, probably no more than a mile to the east of the cemetery.

Mary Ann Erhart had given birth to a new baby girl and her mother, Emma Palmer, was staying at the Erhart home to help her take care of the infant, plus other household chores.

Also at the Erhart home that day were Mary Ann's husband, Neil, their two-year-old son, William, and a niece, Clare Howard.

Because there were no survivors, no one knows what happened to the family that day. Their bodies were found in two places; Mrs. Erhart, Mrs. Palmer and the baby in one place, and the others a little distance away.

The Gathering Fireclouds

The Dawn

It was a blood red sun that rose on September 5, 1881, over Michigan's tortured Thumb District. Months of rainless heat made the forests scorched and brown. A wind was building from the southwest, fanning the heated air like a blowtorch and causing the dried leaves on the aspens and maples to clatter noisily. The pine trees whispered weird noises in reply. Smoke from distant fires scented the air and the animals were stirring. They could smell death in that wind.

The Fury

The black wind came shortly before noon.

Maryia Weitzel and her six children peered out of the door and windows of their wilderness log cabin to watch as the trees whipped and bent in the gale, and the sky turned from deep gray to ebony black. They found they had to light the kerosene lanterns to see.

When she stepped out, Maryia realized that the wind was abnormally hot and the air was filled with whirling black smoke and ash. She knew a forest fire was coming and that the family was in terrible danger.

This was a dilemma. Since moving to their homestead in what later came to be known as Argyle Township, Sanilac County, Maryia and her husband, Paul, had heard stories of a major fire that swept the area in 1871. She knew that people jumped down into wells and streams to save themselves. But Maryia did not think of these things today. Paul was about 60 miles away working as a brick mason and she was alone and frightened.

Her decision was easy to understand. She and the children would walk to the home of George Kroetsch, Maryia's brother, which was about 1-1/2 miles to the northwest. She knew George would know what to do.

There were two ways for the family to get to the Kroetsch home. One was a direct route, through the forest, or she could follow the dirt trails that charted the surveyed section lines. Maryia chose the latter.

No one knows the agony that family went through during those fateful minutes. Maryia and her children, Martin, 14; Margaret, 12; Maryia, 8; Theresa, 6; Julia, 4; and Ambrose, 2, set off walking north on what is now Frieburger Road to Deckerville Road, then west toward the small settlement of Laing at the corner of Deckerville Road and M–19.

As they struggled against the wind and smoke, Maryia, already tired from carrying at least one, if not two of the younger children, must have realized that time was running out. Red firebrands were now flying through the blackness overhead and igniting the woods all around them. She decided to send Martin to the Kroetsch home to get George.

Martin ran that last mile and arrived at the Kroetsch home retching and sick from the smoke. But he told his story and George left in search of the Weitzel family.

Then out of the inky black could be heard a roaring sound unlike the noise of the wind and crackling flames. Dried old tree stumps burst into flame from the intense heat. The wind suddenly shifted to the northeast with such fury that trees were flattened. A red wall of flame came out of the black with great speed and it consumed everything in its path.

Searchers later found George and the Weitzel family on the road where they perished. It was said that George was kneeling in the road with a rosary in his hands.

Martin survived the holocaust.

The Great Fire Storm

The terrible conflagration that swept Argyle Township about 2 P.M. that fateful Monday was part of a five-hour holocaust that left all of Sanilac and Huron Counties and portions of Tuscola and Lapeer Counties in ashes.

The story of the Weitzel family is true, even though the above account is fiction. No one knows exactly why Maryia Weitzel chose to walk such a perilous journey with her six children, or whether her brother George Kroetsch succeeded in reaching the family before the fire overcame them. Three separate stories have been found, and all vary somewhat.

Even though it happened a century ago, the fire is still talked about among the children and grandchildren of the survivors. It was an event so entrenched in the memories of the survivors, it is said that while they lived, they marked

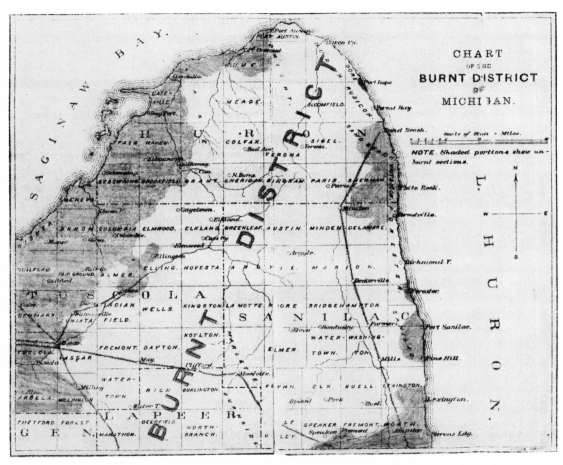

This chart depicts the area ravaged by the Great Fire of 1881. Courtesy of James Lincoln and James Donahue. From Signal Service Notes: Report of the Michigan Forest Fire of 1881 *by William O. Bailey.*

their lives by things that happened before and after the great fire.

The fire burned more than 2,000 square miles of forest, farmland and towns with such fury, and was driven by such powerful winds, that it was said to have overtaken fleeing horses.

An estimated 280 people perished, another 15,000 were left homeless, and countless numbers of domestic and wild animals died. To save themselves, the settlers in this harsh wilderness area climbed down into wells, which in those days were hand dug and relatively shallow, or found cover in streams or open, plowed fields. Those who lived close enough to Lake Huron, ran into the water as the flames bore down on their homes.

Relief to the disaster victims was the first such effort taken on that year by Clara Barton's Red Cross, a fledgling organization that had its beginnings that same year in Dansville, New York. State and local disaster relief organizations also were formed.

From records of that day, it appears that the fire was fanned by erratic gale force winds blowing from the southwest, west, and north. It caused the fire to do strange and unpredictable things. Walls of flames came with great fury upon some people, like it did the Weitzel family, but in other areas the flame burned slowly enough that people were able to fight the fire and save their homes.

The fire was so capricious, it was known to spare a wagon load of dry straw, but burn the same farmer's home and barns. Homes were spared while furniture and bedding, moved to open field for safe keeping, were burned. The fire would split and spare one area, only to return on a shifting wind to consume what it left earlier.

One of the best accounts of the disaster is an 1882 government publication written by William O. Bailey, a sergeant in the U.S. Army Signal Corps, who visited the area in the weeks that followed the catastrophe.

It was Bailey's findings that the four counties were visited by several different forest fires that eventually turned into one single fire. He wrote:

> The course of all the fires, which together made the great conflagration, was mainly toward the northeast. How the fires ran racing over the four counties has been told, but the story gives no picture of the terrible fury of the storm of fire and wind that destroyed the property and lives of the farmers and settlers.
>
> The heat from the flames was so intense that the people felt it while the fires were miles away, and sailors at Forestville felt it uncomfortably at a distance of seven miles. It withered the leaves of trees two miles from its path. While fields of corn, potatoes, onions and other growing vegetables that were not touched by the flames, were roasted by the heat. It even became the cause of an unnatural growth, and peach, apple, and other fruit trees burst forth in blossoms. Fish in the streams were killed by the fierce heat, and after the fires were over, their dead bodies were found floating on the surface of the water. Birds, escaping from these terrible flames, were carried far out into the lake, and, dazed, blinded, and finding no resting place, were drowned.
>
> Several witnesses gave an account of a curious phenomenon illustrating the intensity of the heat. A peculiar blue–white flame would sometimes burst forth from ignited tree stumps, flicker a few seconds, and then the strong winds would extinguish it. It resembled a lighted candle. The phenomenon attracted attention in several places. Even the earth sometimes took fire.
>
> The speed with which the flames and wind traveled, and the tremendous powers they exerted are

almost incalculable. Some of the effects of the wind have already been given. Large boulders were rolled along the ground as if they were pebbles. The conflagration is described as roaring like a tornado, and as giving forth loud explosive sounds, that were terrifying.

As the storm advanced, it uprooted great trees, blew down buildings, carried roofs through the air, lifted men and women from their feet and threw them back violently to the ground, in some cases seriously injuring them. The flames literally raced through the country, licking up villages almost in an instant. . . .

Before the fires came the air was thick with blinding smoke, and the darkness became almost total. In some places it was actually total. Lamps were lighted at midday, and the lights threw shadows as electric lights do. Through the darkness flaming balls of punk fell into villages and fields, and then the fires would burst forth on every side. The flames came rushing on, sometimes in huge, revolving columns, then in detached fragments that were torn by the winds from the mass, and sent flying over the tops of trees for a quarter of a mile to be pushed down to the earth again. Flames were seen to leap many feet higher than tall pines, and everywhere over the burning country sheets of flame were flying in every direction.

The flying sand and smoke blinded people who walked, in the gathering darkness, into firetraps. Those who escaped were blind for weeks, Half naked creatures made their way into village streets, often bearing the charred remains of the dead with them. Many found refuge from the fires in the lake, and even there they were suffocated by the smoke blown from the shores. The cinders, falling in the water, made a lye, so that it was necessary to go down several feet under the surface for drinking water.

Smoke from the fire blackened the skies over Ontario, Canada, causing the people there a great deal of alarm. Some thought it was the end of the world.

The newspaper in Boston, Massachusetts, about 700 miles to the east, reported a strange yellowish color to the sky on September 6. The smoke and ash from the fire actually obscured the sun over Boston for several hours.

Mixing of the Elements

There were numerous natural and man–made elements that combined forces to make the 1881 fire the dreadful killer that it was.

The first people other than the Indians to come to the area were hunters and trappers. But they soon were followed by the lumbermen who saw a potential for wealth in the rich stands of virgin pine and hardwood forests.

Sometime around 1850 lumber camps dotted the shores of Lake Huron. Later they were built inland, mostly along the streams. As the big trees fell, the lumbermen took only the long, straight trunks, leaving piles of tree tops and limbs, or slashings, where they fell. The Thumb was a lumberman's dream. It contained some of the best white pine in the world with trees towering over 100 feet.

The hemlock trees that grew in Forester and Sanilac Townships were cut for the bark, which was used in tanning leather. The lumbermen stripped the bark from these trees and left the slashings, complete with the dead trunk behind.

In October 1871, a fire burned across the forests of Michigan and into the Thumb area. It was remembered as a treetop fire. Like the killer fire of 1881, it was fanned by high winds and it raced at great speed through its destructive course. But the 1871 fire came so fast that it did not consume the trees it killed, but merely left the dead and charred stumps as additional kindling for the conflagration to follow.

The 1871 fire did not burn all of the forests in the area, and lumbering operations continued. But it probably did help hasten the end of the lumbering era. The lumber companies, which owned large tracts of land, had stripped the area of its best timber by 1881. The land was selling for only a few dollars per acre in parcels of 40, 80, and 160 acres to pioneer farmers, many of whom came to the area to escape political unrest in Canada.

Other settlers came from the East, by way of the Erie Canal. Some were Civil War veterans. Polish emigrants established a settlement in Paris Township, Huron County, prior to 1850. Other nationalities included English, Irish, Welsh, Scots, Germans and Scandinavians.

A flourishing fishing industry developed in some of the settlements along the Lake Huron shores, and the land was found to be rich and fertile for farming. It was not long before ships from Detroit and Port Huron were docking at these towns to bring items of trade for the lumber, the lumber products, fish and food produced in the area.

The people who settled inland cut trails through the forest so that ox carts could be used to get their goods to the port towns and to bring home supplies. These trails, many of which have become modern roads, were no more than fire trails through the trees, just wide enough for a wagon. The Thumb area of Michigan was mostly low, flat swampland, and traveling was difficult. It was a common practice to lay logs, side by side, on the roads to buoy up the wagons. These were known as "plank" roads.

The Port Huron and Northwestern Railroad was the first to be constructed in the Thumb. The narrow gauge road came north from Port Huron to Davisville (later Croswell), and continued north through Mills (now Applegate), Farmers (now Carsonville), Deckerville, Minden and Sand Beach (now Harbor Beach), in 1878–79.

When the firestorm came, thousands of farmers only had cleared the stumps and brush of a few acres on each farm. There were stories of how potatoes and other crops were planted between the old stumps. The work of clearing the land was slow and difficult because everything was done by hand. It was common, in fact, for the settlers to burn the old stumps and brush piles.

The air frequently became filled with wood smoke during good dry burning days. George Harper Talbert, a resident of Verona Mills, Huron County, kept a diary in 1877. He wrote on Monday, May 14 that it was a "nice warm day—fires all round—wind go to east awfully smoky."

The general picture of the Thumb area that fateful autumn was one of a disaster waiting to happen. The land was filled with dried trunks and slashings from lumbering operations, charred stumps and logs from the 1871 fire, and good stands of evergreen and hardwood trees regarded then as unsuited for cutting. The newspapers reporting on the fire of 1881 properly referred to the Thumb as a remote wilderness. It also was a dangerous tinderbox.

Interspersed among this forest and combustible rubble were the log cabins of the new pioneer settlers. Beginning the difficult job of clearing the land with picks, shovels, handsaws and matches, they discovered that matches were the easiest tools to use. Consequently, they were seen burning freely.

Weather conditions became the final catalyst for the great fire. Sgt. Bailey described the situation:

In September no penetrating rain had fallen for almost two months. Almost every stream was dry. Many wells had become empty. The swamps had been burned to hard clay by the sun, fiercer in its heat than it had been for years before. The vegetation of the fields and woods had become tinder. The earth was baked and cracked. . . .

The general atmospheric conditions prevailing during the fires were very favorable to their spread. The winds throughout the state, south of the conflagration, were generally from the southwest, while west and north of the firebelt, a disturbing cause of sufficient magnitude produced violent and destructive air currents from the opposite quarter.

Bailey recorded winds of 40 miles per hour blowing from the southwest in Lapeer County:

Conditions north of the burning territory strongly favored north to west winds, and while the fires were raging and generating abnormally high temperatures, abnormally low temperatures prevailed only a few hundred miles to the west and north. It is stated that a cold northerly wind reached the burning district about three in the afternoon.

The mixing of two major weather fronts at the same time the fire was reaching its peak produced catastrophic results. Violent winds were said to come from all directions, knocking down trees, tearing roofs from buildings and flinging people through the air to their death.

One survivor in Sand Beach later wrote to her mother and described what happened as a tornado that came with the fire. Her description may have been quite accurate.

Early Warnings

While the hot sun and dry southwesterly winds worked to prepare the area for the disaster to come, the farmers used the good weather to torch the great piles of dead tree limbs and trunks that kept them from working the soil. Smoke from hundreds of such fires was noticed for weeks.

The earliest recorded problem from these fires came on August 5 when a severe windstorm fanned a fire that had been burning north of Bad Axe into a hot and dangerous inferno.

The *Huron County Tribune*, a Bad Axe newspaper, reported that "when the wind came it swept the fire towards us in a tremendous cloud of smoke and flames that bid fair to devour everything in its path. . . ."

The newspaper account stated that people became excited and ran for their homes. A criminal examination in the courthouse ended abruptly as lawyers, police officers and prisoners quickly vacated the courtroom. Then, at the last moment, the wind changed and the running fire stopped.

In Port Huron, the weather station reported on August 13 that there were "dense clouds of smoke from the forest fires" and that "large forest fires in the west and southwest are plainly visible from the station."

Newspapers throughout the Thumb expressed concern about the fires. The *Lapeer Clarion* reported that "the forest fires are getting to be a source of considerable damage and great anxiety."

On August 30, a newspaper reported that fires in the woods east and west of Capac, in St. Clair County, presented a grand appearance.

Sgt. Bailey stated in his report that a major fire started on Wednesday, August 31, in Lapeer County. This fire, whipped by strong west and southwest winds, advanced north into Marlette, in Sanilac County, then turned northeast and east through Moore, Custer and Watertown Townships.

It claimed two lives in Lapeer County, near Burnside. An account in *Pioneer Families and History of Lapeer County*, said one victim, a Mrs Leonard Leach, put her baby in the arms of a neighbor and joined her husband in an attempt to save their home. Their bodies were found burned almost beyond recognition.

The fire reached the village of Sandusky about 2 P.M. where it destroyed a house, a church and school. From there it traveled northeast raging through the swamps until it reached Deckerville at 4 P.M. where it moved east through the forests of Marion and Bridgehampton Townships.

As the fire worked its way slowly in a northerly direction across central Sanilac County, other towns were threatened. But at dusk, the wind shifted to the north and turned back the flames.

On Sunday, September 4, fires that had been burning in northern Tuscola County had united and were working their way east into Huron County. Small fires and the smouldering rubble of former fires continued to burn throughout the area. By that night the stage was set for the great holocaust to come.

People went to their beds that night, many tired from fighting fires, others from the harvest still in progress. The atmosphere was hot and smoky. The people undoubtedly were concerned about the red glowing skies in the forest around them, but they could not have known of the impending danger.

These were to be the last peaceful hours these rugged settlers were to spend for many days. For some, the night was their last.

The Burning

Tiger! Tiger! burning bright,
In the forests of the night;
What immortal hand or eye
Could frame thy fearful symmetry?

William Blake

It has been said that the fire marched in one great wall of flame, sometimes 100 feet high, across Michigan's Thumb in four hours. That is not entirely accurate.

It is true that some areas were struck by a marching wall of fire, and that there were reports that the fire moved so fast that it outran horses. It also is true that the fire spread from Cass City, where it was recorded between 11 A.M. and 11:45 A.M., to Verona Mills and Sand Beach between 3 P.M. and 4 P.M. the same day.

Whenever anyone remembered to record the time that the fire came upon them, it was sometime within those hours.

This telegram sent on September 7, 1881 reads, "Richmondville in ashes . . . safe down home stay until come or send for you. . . . Will write particulars." Courtesy of Oliver Raymond.

Yet there was at least one report from Buel Township, in southeast Sanilac County, stating that the fire arrived in the evening. And other accounts tell how the fires continued burning and ravaging the area until Wednesday, September 7, when a welcome rain fell and helped put the fire out.

Sgt. Bailey wrote that numerous separate fires were marching that fateful day. He said a fire already burning north of Bad Axe began moving north and northeast from Colfax Township into the townships of Meade and Chandler, where it eventually reached Huron City, Port Austin, Port Hope and Forest Bay.

Another great conflagration originated in Novesta Township, Tuscola County, which burned Cass City and moved east into Greenleaf Township, then grew to be a killer fire that took many lives as it burned through northern Sanilac and southern Huron Counties. It burned through the towns of Holbrook, Tyre, Paris and Minden, and east through Argyle Township.

Bailey said a third conflagration, which started in Millington and Tuscola Townships, Tuscola County, spread into Vassar and Indian Fields Townships. This fire, influenced by southwest and westerly winds, moved east into Sanilac where it joined still a fourth blaze, which had its origins in Lapeer County. This fire moved into Marlette, Omard, Peck, Sandusky and Deckerville.

The heat from the fire was so intense, and it remained to burn everything so completely, that families were reported to have remained in wells and streams for two and three days before thinking it safe to come out. The ground remained so hot that animals burned their feet.

On a Rock

Sometime before noon the mounting firestorm came to the Cass City area. Many people had time to get to the Cass River to save themselves, including Mrs. Cleveland Downing, who lived on the end of Doerr Road, southwest of the village.

As the wind grew in strength, and the flames blackened the air with smoke, Mrs. Downing carried her most valued possession, a sewing machine, with her to the nearby banks of the river. She and a granddaughter spent that terrible day and the following night on a rock in the middle of the river. Also sharing that sanctuary were some wild animals, including a black bear.

The next day, Mrs. Downing discovered that her house was still standing. Her home had been spared.

George Helwig

George W. Helwig was one of the rugged pioneers who settled in the territory near Cass City, moving with his wife, Catherine, and their five children (Nattie, 11; Minnie, 9; Clara, 7; Edward, 4; and Lena, 2) from Lancaster, New York, to the forest 2 1/2 miles east of the village in May 1881.

The family spent the long hot summer cutting and hewing trees with axe and saw, to build their home in a clearing just west of Schwegler Road. A wagon path through the trees was their only link with the nearby village. The Helwigs moved into their home in the late summer, only days before the fire.

When the fire approached, Helwig decided to stay and fight to save the house, instead of escaping to the safety of the nearby Cass River. The house was the product of two months hard labor. His family remained at his side.

A bucket brigade was established. George got up on the roof and the three oldest girls carried buckets of water from the well. Catherine and the younger children spent their time stamping out sparks and throwing water on small fires near the house.

The wind, reaching gale force, was strong enough to knock a man from his feet. Smoke was so dense that breathing was difficult and eyes burned. Temperatures, already reported in the high 90's and undoubtedly made unbearable by the heat of the approaching flames, were just one of the agonies the Helwig family suffered that day.

By some miracle, the house was saved and the family lived to tell their story. After the fire passed, everybody was sick from the smoke. George was nearly blinded and spent several days with moist cloths over his eyes.

An inspection of the property later revealed that while the family was outside battling the fire, an ember had somehow dropped on one of the beds in the house and burned a hole in it. But this little fire burned itself out, unnoticed. The family often remarked that God was watching over them that day.

The house still stands today and is occupied by a grandson and his family.

In the Well

The fire, now gaining in intensity and fanned by even stronger winds, was bearing down on the home of George C. Hartwick, a cooper, who lived just east of the Helwig home on Lamton Road.

Hartwick, 57, sent his family into the hand dug well. Most wells in the area were no more than 20 feet deep and about six feet wide. They usually were lined with field stone. Because of the severe drought, the water in them was not very deep.

Taking refuge in Hartwick's dank well that day were Hartwick, his wife, Harriet, 34, and their five children (Maude, 11; Gilbert, 9; Herbert, 7; Edward, 4; and Isabel, 1).

Before the fire arrived, Hartwick covered the well with some heavy wood timbers. When they caught fire Hartwick threw buckets of well water at them from underneath. Here the family remained for hours, suffering from smoke and lack of oxygen, drenched with buckets of water and ashes from the burning timbers over their heads.

The Hartwicks survived. They crawled out of the well some hours later to find a blackened, still smoldering landscape. Their log cabin was burned to the ground, and Hartwick's tools for barrel making also were lost.

Several weeks after the fire, the Hartwick's pig came home. It remained a mystery just how the animal survived.

Beating Down Flames

As the fire moved east into Greenleaf Township, it approached the home of Alexander MacLellan, located one-half mile north of the old community of New Greenleaf, on Hoadley Road.

MacLellan, who moved to the area with his wife in 1880, also decided to fight to save his home. But he took on the job alone. He sent his wife and their baby daughter, Sarah, into

the well, and used wet gunny sacks and carpet to beat down the flames. The home, a sturdy log cabin, was saved.

Judgment Day

While Alex MacLellan was fighting to save his home, the James Brown and Robert Jackson families, four miles to the south near the town of Wickware, were fighting to stay alive.

Brown and his wife, Mary, and his sister Jane Jackson, with her husband, Robert, like many of their neighbors, came to the area from Canada in 1880 and began homesteading the newly opened territory. They lived close to each other on the Cumber Road in Greenleaf Township. Both families had built a log cabin and had cleared a little land.

When the wind grew that terrible morning and the sky turned black, it threw fear into the neighborhood. The Jacksons gathered at the Brown home to discuss the eerie blackness, which by noon was dark as night. They thought it might be the Judgment Day.

The wind bent the trees to the ground, and everyone had difficulty standing upright. When the firebrands began falling from the sky and fires began breaking out, the families realized they were in trouble.

While the women whisked the children into a nearby cornfield, the men drew a tub of water and carried it with them to the center of the field. They no sooner arrived when the fire was upon them. It came with a roar, with great heat and fierce winds. Huge columns of fire shot out of the black and lit the sky overhead like strange orange lightning bolts. The heat wilted the corn.

Robert Jackson used his black felt hat like a pail to throw

water on the family and keep their clothes from igniting. The firebrands, burning embers carried by the winds, rained down around them.

When the danger passed and the Browns and Jacksons were able to leave the charred cornfield, they found their homes, belongings and oxen destroyed. Only one house remained in the area, and several families were forced to live there for weeks to follow.

The Death Well

There are conflicting stories about the deaths of John and Mary Freiburger and their children. It is generally agreed that the family died of suffocation while packed in a well, but the place of the incident, and even the number of children in the family are unclear.

Lynn Spencer, a rural mail carrier in northwestern Sanilac County, believes the Freiburgers died in the town of Freiburger, located one mile east of M–19 in Austin Township, Sanilac County.

Spencer, an ardent history buff who talked to the older people about the fire, was told that 13 children and their parents died in the well, located near the intersection of Freiburger and Cumber Roads. The Freiburgers and other fire victims were buried on a three-acre tract of land one-half mile south of the town. The land was donated by Louis Peters, a survivor of the 1881 fire.

Arlee Freiburger, of Snover, a nephew who has researched the family history, said he believes the family died on a family farm three miles north of Ubly, in Bingham Township. He has obtained a copy of the land deed and copies of their death records from the Huron County

A speeding fire in Michigan timberlands. Courtesy Michigan History Bureau, Michigan Department of State.

Courthouse to prove his claim. He believes five children died, and that the family lies buried in the Schmitt Corners cemetery, six miles northeast of Ubly.

Arlee Freiburger relates that John was one of five Freiburger brothers who came to Michigan from Waterloo County, Ontario, in the months before the fire. The other brothers located in Austin Township and the town of Freiburger bears the name. It once boasted a hotel, blacksmith shop, Maccabees hall, Lutheran Church and cemetery, Catholic Church, general store, doctors office and a two–room school. The other brothers survived the fire, as did the town of Freiburger, according to his story.

How fickle the threads of historical accuracy! Both stories cannot be true unless two Freiburger families died in wells within miles of one another.

But these two facts are true: the town of Freiburger no longer exists (Spencer claims it burned in the fire); and the Freiburger family is, indeed, buried in the Freiburger cemetery—their headstones mark the spot. The headstones tell us there probably were eight children who died that day. They are arranged in a perfect row, with the first name of each child engraved on the top of each stone: Peter, Christina, Minnie, Frank, Annie, George, Anthony and Mary.

"Run for It!"

Maurice Clifford was a Canadian emigrant homesteading land just east of Freiburger, on the Cumber Road. He had built a one– room log house and barn, and spent the summer of 1881 hewing out logs for a larger, more permanent home. Living with Clifford were his parents, his wife and their four children.

Ellen Ryan, of Capac, Michigan, Clifford's granddaughter, has written about the way her family survived the fire:

> Grandfather Clifford had cut and hewed logs for the new house. With grandmother's help they cleared some land and piled the logs in the yard awaiting the house raising.
>
> On the day of the fire the atmosphere took on a hazy, gray appearance. Then smoke rose in the west and wagons began to pass the farm. Some who knew my grandfather called out, "Come on, Maurice, make a run for the lake. All the land west of here is on fire!"
>
> The fear was evident as they lashed at their horses to speed them on.
>
> My grandparents decided to stay, and they set up a sort of bivouac near the stone well. Grandmother Clifford lay quilts on the ground surrounding the well. Then the ground and blankets were soaked. Here the whole family spent the night.
>
> All through that terrible night my grandfather poured water on the ground and blankets. Because of the intense heat the small area dried quickly. All except my grandfather Clifford lay beneath the covers.
>
> The fire was more intense because of the pile of logs. The night was a long and dreadful ordeal for every living thing. Wild animals came up and dashed away in search of safe shelter and water.
>
> By the time morning came, an eerie silence came over the area. You could hear an occasional snapping and cracking from a burning stump or the wood pile.
>
> My grandfather temporarily lost his eyesight. Facing the smoke and burning cinders burned his eyes and he had to be led around for several weeks.
>
> Grandmother Clifford picked her steps to what was once the house and gathered a fist full of salt. Then she went to the garden and dug the baked potatoes from the hot earth. Potatoes and water tasted good after the long anxious hours.

Maurice Walsh

This story about an Austin Township man who escaped the fire but lost his family comes in different versions. Because the stories have been handed down through the sons and grandsons, there has developed margin for error in the telling.

Maurice Walsh and his son were in the forest early in the morning to hunt. They left their log cabin home—about one mile east of Freiburger on the Cumber Road—at dawn because they knew the day promised to be hot.

As they tracked through the thick brush, much of it regrowth in areas burned by the forest fire of 1871, they were surprised by two deer that crashed unexpectedly out of the trees, passing them in great leaps. The animals came upon Walsh so fast he had no time to raise his rifle.

It was not long before Walsh noticed more animals, both large and small, moving in an easterly direction through the trees. They seemed more frightened of whatever it was they were running from than of Walsh. He was a good woodsman, and he knew something was wrong.

Because they were sheltered by the trees, Walsh and his son were not conscious of the building wind until the trees began to thrash. When the sky grew dark, and an ominous smell of woodsmoke filled the air, they decided to turn for home.

They had several miles to travel and it became a long and fearful walk. Walsh knew that a forest fire was bearing down on them. The day grew like night, until they found it difficult to see their way among the whipping trees. Before they reached the cabin, firebrands of burning ash were falling around them, setting small fires everywhere. They

were desperate by now; they knew they could not get caught in the woods and survive. Walsh also was concerned about reaching his wife and two children at the cabin.

At last the two came coughing and stumbling out of the burning woods. The cabin was there, just as he left it, but the wind was blowing burning embers onto the wood roof shingles and they were starting to burn. There was a roar in the forest to the west.

Walsh found the door to the cabin open but nobody inside. After a quick search of the home, he grabbed his boy and together they climbed down into the well. A herd of cattle, fleeing the fire, stampeded through the yard and past the well.

While the father and son huddled together in the cold water, the great fire ravaged the world above their heads. By now the fire was burning so hot that it sucked the air from well. For anxious minutes, Walsh and his terrified son found themselves in a vacuum, unable to breathe. Then, at last, when their lungs could draw, the air that came in was hot and burning.

Burning embers fell on them through the open well and Walsh splashed water on himself and his son. They remained in the well for hours, while above they could hear the sounds of crackling flames destroying the trees and their home.

When at last they could emerge safely, Walsh began a search for the rest of his family. He found their charred remains not far away, where the fire had overtaken them.

Tobacco Leaves

While Maurice Walsh was losing his family to the fire, William D. White and his wife, Charlotte, were saving themselves and their six children in a field of green tobacco, just four miles north of Tyre.

Mrs. Florence Peters wrote the story just as her mother, one of the six White children, told it to her many years ago:

> It got so dark at 11 A.M. that grandma had to light the kerosene lamps to see to eat their noon meal. The children wondered what was happening. My grandfather knew and started to plan how to keep his family alive.
>
> The family had planted a field of tobacco, which had wide green leaves.
>
> My mother said that grampa's field of tobacco probably saved their lives. In the middle of it, bedding, some clothing and every available pail or bucket of water was taken, along with mops and brooms. All this time her father was working with a wet towel over his face because the smoke was so bad. Then he took the family out there, made a bed of bedding and made them all lie down and cover their heads.
>
> He turned his cattle loose and drove them towards the river out behind the farm. Chickens and pigs were left to fend for themselves. He had children to care for.
>
> He and grandma and a single neighbor man on the next place gathered to fight the fire. My mother said it was so hot she nearly roasted under the bedding, but if she stuck her head out she was hit with a wet mop or broom. The falling cinders or firebrands burnt many spots in the quilts. But the parents put the fires out and kept the quilts wet.
>
> All that afternoon and night they stayed there. All their buildings burned. The wind blew firebrands for miles and set new fires. Everybody had to keep something wet on to keep from being set afire themselves.
>
> So the next morning they all took what they could carry and started down a hard baked road, to see who else they could find alive. All was mostly burned to ashes and there were no people around. They kept on walking until

they came to a farm where people by the name of Clelands lived.

> The Clelands had been summer fallowing to kill witch grass after their crops were off, so had most of their land plowed up. When they saw the fire coming they plowed more close to their buildings and pumped water in every container. Their buildings were standing, but they fought all night to keep fire off their roofs.
>
> The sun came out in a thick grey fog. Many people did like my grandparents. They went down the road hollering at every place, but all gone.

The Wedding Dress

The Peter Sparling family lived just north of Tyre, across the Huron County line. When the fire came, Sparling lowered his wife down into the well with her most cherished possession, her wedding dress. The family survived the fire, and the wedding dress still is cherished by the grandchildren, who know the story.

Sparling, a big and rugged Irishman, came to Tyre from Canada prior to the fire of 1871. He arrived by boat with tools and oxen at Forestville, then walked 18 miles through the forest to the land he chose to homestead.

Sparling cleared the land, built a home, and carried a cast iron stove on his back from Forestville, before bringing his family into the wilderness.

The Sparling family survived the fire of 1871 and knew what to do when the 1881 fire was bearing down on them. They moved into the well, where Peter had installed cross pieces of wood, and lowered straw ticks, or mattresses for the family to lie on. He also put cross pieces of wood over the top of the well, and wet blankets over them.

Sparling's son, "Little Pete" Sparling, 23, was busy that

day clearing some neighboring land for himself when the fire caught him. He put wet blankets over the oxen and himself and laid near a stonepile. Sparling used a plate to shield his face as the fire passed over him. He survived.

The Burning of Tyre

The community of Tyre, which bears the name of the ancient city of the Phoenician Empire, still exists at the northern edge of Sanilac County.

Nestled on an angling dirt road, it is a small community of two or three businesses and a smattering of homes.

In 1881, Tyre was a more important place. It was located at the junction of two well-traveled roads: the old State Road, which angled directly across the Thumb in a northeasterly route; and the Pioneer Trail, which went west from the Forestville docks. The State Road was abandoned many years ago, although its bed can still be seen from the air. The Pioneer Trail, later the Bay City–Forestville Road, was rerouted to follow surveyed section lines that pass one-half mile south of Tyre.

Lynn Spencer said Tyre had 14 businesses including a mill and a doctor's office in 1881. A wood boardwalk was built along the west side of Main Street.

The fire left the community in ashes; nearly all of the buildings were destroyed. Surprisingly, there was little, if any, loss of life.

Doris M. Norris, of Yale, Mich., whose grandparents were among the Tyre survivors, said that Alfred Gunney lost his store when the fire came: "All that was left was the iron safe and it was blistered from the fire. Later my grandfather used this safe in his elevator at Tyre." The safe still remains

in use today in a Durand Furniture store, operated by a cousin.

The *Port Huron Times Herald* reported on September 12, 1881, that the Tyre Post Office, store and attached home of owner Amelia O'Sullivan, and some nearby barns, also were lost. But the school, located just east of Tyre in Paris Township, survived the fire and many families stayed there.

Riding into Hell

George Gillespie, a resident of Tyre, actually rode a wagon pulled by a team of horses into the heart of the inferno and lived to tell the story in the *Port Huron Daily Times* on September 9, 1881.

He was in the village of Minden when the fire approached from the west. He decided to try to get home to his wife and children, five miles to the west.

"He started to drive home but it became so dark that the horses could not keep on the road. Finally a lantern was procured which he fastened to the pole between the horses," the paper reported.

Gillespie drove his team into a forest that was burning on both sides of the road and nearly managed to get to Tyre before the firestorm forced him to stop. He said a church located close to the road was burning and the heat prevented the horses from passing.

He then abandoned the horses and ran around behind the church, where the land had been cleared. When he reached the village not a standing thing was to be seen. His family had either left the village or had perished.

After hunting around for a few minutes he spied what appeared to be the form of a woman. He found it

The September 24, 1881 issue of Frank Leslie's Illustrated Newspaper *depicts a family fleeing from the advancing flames of the fire. Courtesy American Red Cross.*

was a neighbor, Mrs. O'Sullivan. She was almost dead from suffocation. He got an old kettle and started to a well 40 rods off to get her some water. On the way he met a man who told him that his wife and boys were hidden behind a stone pile some distance off, where he discovered them. His wife was suffering from the smoke and fire which she inhaled, but fortunately escaped without being burned.

Mr. Gillespie found the bodies of two Polish residents from Paris Township who succumbed to the flames. Other bodies were lying in the road and fields.

Gillespie transported his wife, who remained ill for some time after the fire, to Port Huron with the family where she was given medical treatment. They went from Forestville to Port Huron on the steamer *Saginaw* and Captain George Tebo took up a collection among the passengers. He raised $22.30 for the Gillespie family.

The Ubly Chronicle

The late Robert J. Hagen, who witnessed the fire while he was an eight-year-old boy living on his father's farm near Ubly, wrote one of the few eyewitness accounts of the fire. The story was published in the *Port Huron Times Herald*, on April 29, 1956.

Unlike the other fire swept areas, there apparently was no wall of flame advancing at high speed upon the Hagen farm. Instead, Hagen wrote, farmers in the neighborhood had time to band together in cooperative efforts to save some farm buildings.

The woods were smoky, but Hagen recalled this was not alarming because it was common for the farmers to be burning to clear their land. But Hagen's father, also named Robert, was called midmorning that fateful day to the nearby farm of Alex Donaldson, a brother-in-law, because a fire in the woods was threatening the Donaldson barn.

Their efforts were in vain, however. As the wind grew, both Robert Hagen and Alex Donaldson became more concerned about saving themselves rather than the barn.

Hagen's sister, Sarah, and brother, George, reported that a wood rail fence next to the woods on the north edge of the Hagen farm was on fire:

> Mother hurried over there and thought she could save the rest of the fence by throwing down rails at each side of the fire. She started to do that when she saw the fence was on fire farther along. I had to run along with her. . . .
> As we hurried back to the house, the woods to the west were on fire and the burning leaves were falling like a snowstorm but it was fire instead of snow.

As the fire advanced on the Hagen home, Hagen's father arrived home and herded the family out of the log cabin and into the milk house, which was dug in the side of a clay hill. Farmers call them root cellars.

The family put bedding on the floor and 11 people crammed themselves into the small underground room—Hagen, his six brothers and sisters, his parents and grandparents.

> The fire seemed to roll in from the woods and the strawstack, barn and house were all in ablaze at once. Father went back to the house to try to save the sewing machine. He got it out into the garden, but it burned there.

He burned his hands quite badly and breathed in so much smoke and fire that he was in poor health for a long time.

After we were all in the milk house, father opened a door in the east end, right at the top, to let out the smoke. Then he and George sat with their backs to the door. Every little while, father would step outside and, if the ends of any of the logs were on fire, he would rub it out with dirt.

I went over and sat down with my back to the door but it was too hot for me. So I went back where it was more comfortable and lay down on the floor and went to sleep.

The Hagens remained in the milk house all night. They could hear the horses running back and forth in the pasture, whinnying. The horses survived the fire, but their feet were so burned that their hooves came off. But they grew new hooves and were good horses afterwards, Hagen reminisced.

The cows broke out of the pasture and ran to the nearby Cass River. "We had a pet cow we called Blackie. She ran into the river and laid down in the water near Mr. and Mrs. John Dunlap and their three children, John Jr., Milo and Ellen. Mr. Dunlap noticed a large tree on the bank that looked as if it was going to fall right on Blackie, as the wind was so strong. He spoke to Blackie and she got up and moved down the river beyond them and laid down in the water again. Shortly afterward, the tree did fall, right where Blackie had been lying."

Three neighbors on the Thomas Barnes farm, located south of Hagens, died in the fire. Hagen wrote that Barnes and his son, Charley, were away on a thrashing crew, but his wife was home with their infant daughter, her invalid mother and an aunt.

When the fire came, Mrs. Barnes picked up the little girl and started for the river a short distance away. When she was about half way there, she found the little girl had died in her arms. She laid her down and ran to the river, her clothes on fire. She plunged into the water where she was found next day. She was badly burned but recovered. Her mother and aunt both perished in the home.

Minden Terrors

The great fire was bearing down on the community of Minden, five miles east of Tyre in Minden Township, with hurricane winds. The people thought the town was lost. At the last minute, however, the wind shifted to the north and the fire was turned enough that a large portion of the town was saved. But the people of Minden were not spared the terrors of the fire. As J. H. Shults, editor of the *Minden Post*, wrote:

Monday morning dawned smoky and dismal, and by 11:30 A.M. lamps were lighted in all the houses. At noon, Egyptian darkness prevailed. It was darker than the darkest night we ever saw. Objects could not be distinguished a foot distant.

As the fire neared the village, the atmosphere became lurid, seeming to be filled with flames. A strong wind had been blowing from the southwest, which increased by 3 o'clock to almost a hurricane. The air became so hot that leaves on the trees were cooked while the fire was yet a mile distant. Buildings equally distant became so hot as to burn the hand when touched.

All communications with the country west and north was cut off, and many who were in Minden on business vainly attempted to return to protect their property from the fiery monster.

Strong-hearted men were in despair, women were frantic, many fainting, and others imploring their protectors to take them to the shore as the only place of safety. A number started for Forestville (seven miles to the east) but when they reached Charleston, that village was in a mass of flames, and they were compelled to return and proceed to White Rock.

Most of the people remained in Minden, however, and fought the flames with the determination of those who struggle for life. Sheets of flames were flying in every direction, and incipient fires were extinguished in all parts of the village. At last, when the fire had reached the western outskirts of Minden, there came a lull in the hurricane, and the wind commenced blowing to the northward.

In ten minutes Charles E. Snyder's house, north of the village, was in flames, and George Grummett's house and barn immediately followed. The fire south of the village was driven northward to C. A. Ward's elevator, which was quickly consumed, and but for the heroic efforts of R. D. O'Keefe the depot would have burned also.

A report in the *Port Huron Daily Times* on September 12, 1881, told of a strange incident at the home of Thomas Brocklesby, located opposite the mill.

As the fire was bearing down on the house, Brocklesby gave up all hope of saving the buildings and moved some bedding and other cherished possessions into a nearby orchard where he thought they would be safe. But the fire drove through the orchard, causing Brocklesby and his family to flee to a potato patch, where they buried themselves in dirt. Afterwards they discovered their possessions in the orchard burned, as were their barns and fences, but the house was spared.

Shults wrote that the rest of the village escaped the fire:

By six o'clock people began coming into Minden from the west, having barely escaped with their lives, and when morning arrived hundreds had found their way here in a half nude condition, burned and blinded by the smoke. We then began to have a faint realization of the extent of the fire. Then we first understood that not property alone, but human lives had been swallowed up.

To the west and north the roads were lined with the carcasses of horses, cattle, sheep, swine and poultry, cooked and charred almost to a crisp. Then human beings, alike burned and charred, were found. Some were still alive, with their feet, hands and face literally baked. Some had their ears and noses burned off, and their eyes almost burned out of their sockets.

The sufferers kept pouring into Minden, and our village was turned into a general hospital and boarding house, with all of the generous citizens acting in the capacity of unpaid keepers and attendants.

The Panic

The people of Minden who chose to flee the village and try to reach the safety of Lake Huron, seven miles to the east, became caught in a battle for their lives. John Stanton, a reporter for a Detroit newspaper, recalled the story in 1908:

Minden people fled pell-mell in the direction of the lake shore, some in buggies, some in wagons, some with wheelbarrows, baby carriages and hand baskets, some laden with valuables and some with junk, goods wildly, hastily grasped when panic fell on the town as the bombardment of the cinders began.

It was a frightened, motley crowd that left homes and stores and saloons unlocked and forgot to take its bread and provisions as it fled, but remembered to carry off an old clock, a cracked teapot, a feather pillow or the like.

As the mob journeyed on it came to a patch of forest within a mile of the little village of Charleston, just half way to the lake. The roadway was made partly of logs. These were blazing. The woods were flaming. Deserted Charleston was burning. The fire had outrun and headed off the Minden fugitives.

So they turned into a plowed field by the roadside and prepared to do battle as best they might for their lives. Flames might not reach them there, but smoke might choke them. So they overturned wagons and buggies and with horse blankets made tents. Under these they put the aged, the children and the women.

Then all night long, from a spring at the back of Kelly's field, men carried water and kept the improvised refuges as damp and as safe as possible.

So terrible was the smoke that men often had to lie face down in the field for a time to get respite after carrying water. The passing of time could not be told. There was no moon, no stars— nothing but smoke and flame.

"I don't believe this is any forest fire," said "Yankee" Smith. "This is Judgment Day and the world is burning up! God forgive me for my sins!"

Finally there came a morning—such a dismal morning! There was no rosy, dewy dawn—no peeping of the sun above the banks of the haze. The watches said it was morning. Eyes swollen and smarting, painfully could tell in spite of the still dense smoke that somewhere it was daylight.

The company in the field was hungry and weary. A detachment set out to see if anything remained of Minden—and soon came shouting back to say that the town still stood.

Charleston Destroyed

Ted Schubel, a lifetime resident of the Charleston area, said his father, Albert Schubel, and grandfather, Gottlieb Schubel, battled the fire on their farm on Schock Road, 1 1/2 miles northeast of Charleston.

The Schubel family saved the house and barn by carrying buckets of water to wet the roofs. Cleared fields surrounding the buildings helped keep the fire from reaching the buildings, but Schubel said large balls of fire, mostly wood shingles from the burning town of Charleston, were falling on them from the sky.

The town of Charleston, which earlier was known as Cato, was completely destroyed. It was a town of about 35 buildings, including two blacksmith shops, stores and a grist mill.

The *Port Huron Daily Times* reported on September 7 that Charleston was consumed by the fire within 20 minutes, and that not even a fence post was left standing: "At this place an old man named Henry Cole refused to leave his house and was burned in it. Mr. Palms was also burned to death."

Lost Shoes

Life for the early settlers was so severe and basic commodities so difficult to replace, people chose to save from the fire what we might consider unusual things.

As the great firestorm came upon the home of Julius and Henrietta Stroschein, just south of Charleston on Hunt Road, Mrs. Stroschein decided to make a special effort to save the shoes of her three children, Augusta, Fred and John.

She ordered the children to put their shoes in the boiler, a large copper tub used for washing clothes. The two older children were then told to carry the boiler with them to a nearby creek.

As they ran, probably frightened by the black wind and

falling firebrands, the boys dropped the boiler and its cargo of shoes. They ran on to the creek where they survived the fire. The shoes burned.

Perishing Poles

As the fire began to sweep over Paris Township, in southern Huron County, the powerful southwesterly winds met a cold front bearing even stronger northwest winds. The slamming together of these two weather fronts affected the fires in strange ways.

Sgt. Bailey's report stated that the fires came into Paris Township at 2:20 P.M. from the west as a solid wall of flame 50 to 100 feet high. When the northwesterly winds hit, things started to happen. A 1,000-pound wagon was picked up in nearby Sherman Township and hurled 15 rods through the air. Trees were flattened, roofs were torn from buildings, and the fires began an erratic dance of death.

And yet, while Paris Township became the deadliest place in the fire zone that day, many miraculously escaped the fire.

Bailey wrote, "In the midst of the general destruction a shanty would be left untouched, or a gate would remain intact while the house, barns, granaries and fence were destroyed. Northeast of Parisville the fires divided, and the property of some farmers was burned while that of others was untouched."

A wooden crucifix, erected by the Lemanski family, still stands today at a place where the fire divided. The family members were on their knees that day, asking God for deliverance and their farm was spared. The shrine was erected as an expression of gratitude.

Twenty-two people died in Paris Township in the Fire of 1881. However, as the capricious fire came to the road east of Parisville, it failed to cross the road for an area of several miles. The people in the unburned area saw their deliverance from the fire as a miracle and erected a crucifix as a memorial. In 1983, the crucifix was moved back from the roadside and enclosed.

Bailey reported other miracles in Paris that afternoon: "At the Polish church in this village, a cluster of seven houses within an area 80 rods in diameter escaped, while everything outside this area was burned. On section 17, a rick of hay and a surrounding rail fence escaped destruction while the owner's house, barn, stable and fences were entirely destroyed."

Although it bears the name of a famous French city, Parisville was known as a Polish settlement. The people, like others in Michigan's Thumb, were emigrants who chose to live together for ethnic and language reasons. They were hard-working farmers and devout Roman Catholics.

The stories of how they died are fragmented, probably because there were few survivors. It was reported that 17 people died in a field north of the village.

One account, written in 1884, told how Father Joseph Gratza, pastor of St. Mary's Parish, nearly died in the fire. The story, published in *Portrait and Biographical Album of Sanilac County*, said Father Gratza became alarmed by the approaching forest fire and began ringing the church bell.

While he was inside the church, the fire engulfed the building, causing the bell tower to collapse: "That the priest was able to escape is almost miraculous. Some people racing away in a wagon saw their beloved pastor standing stupified in the parish cemetery. Hurriedly they threw the almost unconscious form into their wagon and sped away to safety."

Father Gratza was one of the area representatives who traveled to Detroit the next Wednesday to solicit help for his stricken people. He reported that 110 families in Paris Township were without food and shelter, and that 28 bodies had been found.

The flames came with such speed that many people died where they stood. Josephine Loch and her five children attempted to dash a few hundred paces to the farm of her parents, but they were caught in the fire and burned alive. Apparently, her husband Frank, survived.

Others listed among the dead at Parisville were Theresa Sperkowski, her five children, Maryana Gura, Leo Kubecki, Theofila Danielski, Mathew Nielewski, Mary Nierzwicki, Victoria Mura and Anna Wrobel.

The *Port Huron Daily Times* on September 9 printed the following list of others who died: Mary Zybeski, Mr. and Mrs. Finlay McPherson, Mr. and Mrs. James Spirkowskie and their five children, and Mathias Naleski, who was smothered while hiding in a well.

The carnage left in the wake of the fire was described in a *Daily Times* interview with George McDonald, of Minden, who drove a horse and wagon north through Paris Township on Tuesday, September 6.

Bodies are lying in every direction. Four miles west of Minden Mr. McDonald came upon the ruins of a house that had been occupied by a Bohemian family named Weisenbugher. Near the foundation lay the father, mother and two children, all burned in a horrible manner. The mother was partly delivered of an infant when found. . . .

They came across a woman who had every vestige of clothing burned off her. Other bodies were found but could not be recognized. Mr. McDonald says that the newspaper reports have not told half of the terrible story, and that language fails to depict the awful state of the people in the burned district.

Ira Humphrey

Ira Humphrey was a U.S. postman who died in the fire as it raced northeast through Marlette and Lamotte Townships.

Numerous stories have been told about Humphrey, and his death is mentioned in a poem on pp. xvii–xix. The man has become legendary. The stories about him vary so much that getting to the truth after 100 years has become impossible. The man was either a fool or a hero, depending upon the story. The following story depicts Humphrey as a hero.

Ira Humphrey took his job as a mail carrier seriously. At age 50, Humphrey prided himself on never missing his appointed rounds to the postal stops that included the little towns in the southwest corner of Sanilac County.

There was no rural mail delivery to farm homes in 1881. Rather, the mail was delivered to towns and the townspeople and farmers picked the mail up at a rural store which also served as a Post Office. Humphrey carried mail packets to towns not served by the railroad. He was, in effect, a one-man pony express.

On the morning of September 5, Humphrey kissed his wife Martha goodby, and saddled up his horse for the daily ride from Marlette through the country trails.

His wife and son were concerned about this day's trip, however. The wind was up and the sky was filled with smoke. A similar wind only last Wednesday had started a terrible fire that threatened Marlette and made the people in the rural areas run to the creeks and fields to save themselves. They begged him to stay home.

Humphrey ignored their pleas and rode his horse into town where he filled his saddle bags with the day's delivery. While at his task, people in Marlette told of ominous stories about a terrible forest fire that was raging in the country to the south and west. They advised Humphrey to stay home that day with his family, at least until the wind died down.

The moment of despair as a family realized the total destruction of their world about them is depicted in this issue of Frank Leslie's Illustrated Newspaper. *Courtesy American Red Cross.*

"My horse can outrun any fire that ever burned," he boasted.

Little did the man know that this fire, fanned by winds strong enough to level trees and rip roofs from buildings, would generate a wall of flame that would outrun the fastest horse. When he rode north out of Marlette that day, Ira Humphrey was riding straight into trouble.

Ira threw the leathery mail pouches across his horse, mounted and rode off down a wooded country trail. Within an hour, he and his galloping horse had death for a companion. The great fire was bearing down on his flank and trees were bursting into flame on his left and right as he rode. Hot firebrands fell from the sky.

Ira knew that a mile ahead there was an open field where he and his horse could find safety from the fire. He believed the odds were still on his side.

As he rounded a curve in the trail, Humphrey came upon an exhausted woman and her terrified daughter. They were on foot, and moving as fast as their feet could carry them in the same direction Humphrey was going. They were partially wrapped in wet blankets.

Humphrey knew the territory and he knew these people were doomed unless he helped them. He dismounted and helped the woman into the saddle. He then lifted her daughter and sat her behind her mother. Humphrey sent the horse galloping on down the road to safety.

The woman and her daughter managed to reach the open field in time to save themselves. The horse, with a badly burned mane and tail, also survived.

Humphrey's charred body was found on the road, not far from where he made his gallant sacrifice for two people he did not know. They found him slumped over the mail pouch, as if he was making a last effort to save the mail it contained.

Race to the River

As the fire roared east into Moore Township, the people fled their forest homes. Those who lived near the Cass River went there.

Among the last to arrive at the water that day was Robert M. Moore, a remarkable fifteen-year-old boy who had been working to clear family property about one mile east of the river, on what now is called Moore Road.

With the fire approaching, Moore loaded a wagon with bedding and barrels of water, and with his frightened team of horses, made a desperate run toward the advancing fire. He knew he had to reach the river or perish.

It took courage for a boy of 15 to drive a team of horses into fire. Burning firebrands, carried by the wind, fell around him starting small fires. Burning tree limbs crashed to the road. The fire burned the hair off the horses' legs. They became wild with fright. To calm them, Moore rode on the wagon tongue, holding on to the harness. He also threw wet blankets over the horses and talked them forward until at last they reached the river. Moore drove the horses, still hitched to the wagon, right into the water where he joined many other people, horses, wagons and cattle.

The Rescue

William F. Smith, one of the early pioneers in the Sandusky area, survived the fire. After the fire passed and he could travel, he went to his brother's home a few miles away to see about his welfare.

Flying embers were still falling, but the worst of the fire had passed.

When Smith approached the cabin he was relieved to see that it was intact, but he was bothered by an unusual stillness over the area. Dead pigs, chickens and other animals littered the yard. There was no sign of life.

When Smith entered the cabin, he found his brother, and his brother's wife and their two children, lying unconscious on the floor.

While dragging the people out of the smoke–filled home and reviving them, a falling ember set fire to the cabin. Smith used milk from the milk house to put out the fire.

Narrow Escape

As the fire raced its deadly course through Custer and Wheatland Townships, it nearly claimed the lives of thirty-year-old Jessie Boice, and her two-month-old son, Albert.

Jessie, the wife of Hamilton Boice of Deckerville, was alone with her baby when the fire roared down on her home. Like so many other people, she waited too long before leaving the home to seek safety.

She ran with the baby in her arms while the supercharged atmosphere overhead made the big hemlock trees explode in flames. Sparks flew from the treetops and the dry parched grass burned under feet. She knew death was imminent. She remembered an old, abandoned well nearby

and without a thought, jumped into the hole and landed on soft, damp dirt about six feet down.

Jessie Boice and her baby survived.

Red Sky over Downington

Matilda Irwin, one of the survivors of the fire, told of the things she saw while a girl of 11 on her father's farm at Downington, on the Marion and Bridgehampton Township line.

She recalled a great cloud of smoke and running to her father, Andrew Phillips, to find out what was burning. He was working with a team of oxen clearing land about a half mile from the Phillips' home. Together they hurried home.

> My father hooked the oxen to a plow and plowed several furrows between our property and the nearest woods in the direction of the fire. My mother took us children into the root cellar, which was dug into the side of a hill and covered with earth. She brought a rocking chair and some food and told me to take care of my four-month-old brother.
>
> The sky had turned to red like a red hot kettle turned over our heads. I thought the world was coming to an end. The sun was purple from the shadow of the smoke.

"Pull the Pin!"

Harvey H. Krohn of Port Huron tells an interesting story as it was once told to him by a Mr. Fogel, a former fireman on the Port Huron and Northwestern Railroad.

According to Krohn, Fogel was working on the northbound train as it rolled into Deckerville on the day of the fire. The fire swept in from the southwest, ashes and firebrands falling on the train. It was obvious that it would only be a matter of minutes before the train was engulfed.

At this point the conductor and the engineer became embroiled in a heated argument. The conductor objected to the engineer's plan to take the train back to Port Huron, 60 miles to the south. He said their orders were to "go up today and back tomorrow, and orders are to be obeyed."

The engineer, obviously unwilling to sacrifice himself for the sake of an order, thought the best chance was to make a run for Port Huron. The conductor said the train would remain in Deckerville.

As the fire leaped closer, the engineer ordered Fogel to pull the pin separating the engine from the cars. They built up a head of steam and started a wild ride through the fire and smoke to Port Huron.

The conductor ran down the track after them, shouting and shaking his fist in the air. He survived the fire by running to an opened plowed field.

Fogel and the engineer got to Port Huron safely, after some hair raising passages through burning forests and over bridges threatened by the fire.

Train company officials could not actually discipline the two because they did bring the engine back out of the fire. Yet they could not be rewarded because they did disobey company orders.

One note of irony: Had the engineer followed orders and made the run northeast to Sand Beach (now called Harbor Beach) as planned, the entire train might have been saved. Sand Beach was threatened by fire, but the village did not burn.

Burning Bad Axe

The fire came upon the Village of Bad Axe, in the heart of Huron County, with such speed that many people ran for their lives.

The *Huron County Tribune* reported:

> The fire burned upon us as if the atmosphere had been turned into flame. It seemed but a few minutes from the time everybody was busy in their shops and offices until almost the whole village was in flames. It certainly could not exceed half an hour. No one had time to remove anything and the inhabitants of over fifty dwellings barely escaped—some to the court house, and others who could not reach that building out of the village to the east.
>
> Those who took refuge in the court house, by a heroic effort, saved that building and their own lives by an all night fight with the fire fiend. Those who went east passed through a cloud of fire until they came to the first considerable opening, where they dug a large trench, into which they put the women and children, covered it nearly over with boards and wet blankets; the men by turns keeping watch until the fire in its march went by. For a little time it seemed impossible that any should survive, the heat became so intense.

Chet Hey, who for years served as historian of Bad Axe, talked to survivors of the fire. He once wrote the following vivid account of the scene at the court house:

> About 1 P.M. the wind became a gale. Smoke, sparks and even burning brands seemed to fill the air. . . . Fear and panic came to every heart. Most persons rushed to the court house, recognizing it as the one possible source of safety. It was the only brick building in the town. It was soon crowded with men, women and children, about 450 persons.

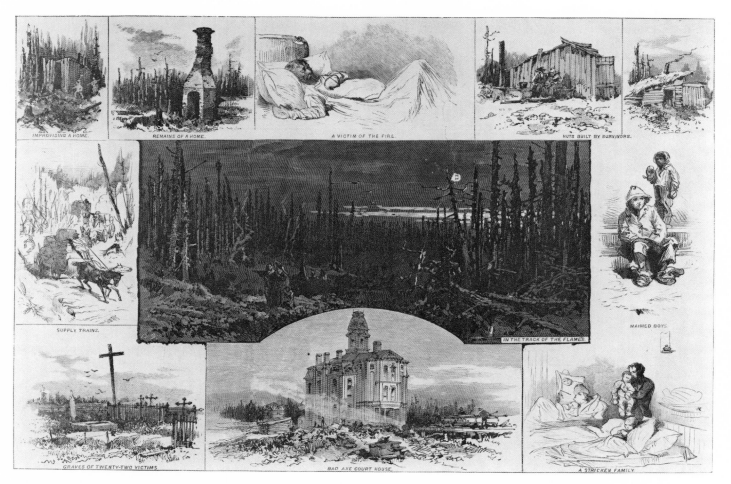

Harpers Weekly *of October 1, 1881 shows scenes from the Great Fire of 1881. Courtesy American Red Cross.*

In the court house, filled to overflowing, were weeping women, crying children, and grim faced men. The well fortunately was on the east side of the court house, which protected it from smoke and wind. Providently it did not go dry. Continuous relays of men pumped, many blinded by the smoke. Others carried the water to protect the building, which housed their families.

Peter Richardson remembers H. G. Snover, J. M. Cary and L. H. Durfy pumped until worn out and that he and others carrying pails of water around the south end of the court house had to have other men go behind and push them against that terrible wind.

Soon the smoke was so dense that one could not see any distance; breathing was difficult. Exhausted men had to be helped into the building. Bandaged eyes were common.

When the lull came, before sundown, with all the buildings of the little town burned, but an occasional scattering home and the court house, which had been the refuge, each had lost his all. Many had hardly clothing enough to cover them. There was no food, no animals, seemingly no future. But there had been no casualties here. All had survived.

That night, it was a sad group in a sorry plight. The oak floors were not comfortable beds, but were all that could be had.

Mrs. Lorna Heath, a Port Austin elementary teacher, also described the Huron County fire in a paper now filed in the Bad Axe Public Library: "The fire did not enter Bad Axe in a direct line, but circled high in the air over the tops of buildings."

Leaping Balls of Fire

From Bad Axe, the fire travelled east through Verona toward Sand Beach at such speed it actually jumped in large balls of flame from point to point.

Roderick Park, a Bad Axe man who was trapped in the fire north of town and survived by climbing into a well, wrote in his article, "The Thumb Fire of 1881,"

> From Bad Axe eastward, the onrushing flames would leap high into the air, then descend to the ground like a bouncing ball, burn everything before them, then rise for another leap.
>
> Under the arch made by leaping flames, the area was unscathed, thus many buildings in the path of the fire were not burned. Where they struck there was utter destruction. Farm implements had every particle of wood burned off. Handles were burned from the plows left in the furrows. All that was found of wooden harrows was the iron teeth. Any tree or fence or house or barn struck by the descending flames withered into nothingness, and this destruction went on until stopped by the waters of Lake Huron.

The *Mayville Monitor* explained the phenomenon of leaping flames: "Any fire of major proportions creates its own draft. That is, it must suck in huge quantities of oxygen to continue its force. This . . . accounts for the fact that the fire actually travelled for long distances in the air, leaving large unburned areas before it would settle to the ground again to gather new and greater force."

The large jumps made by the fire would explain why several people managed to travel the roads between Bad Axe, Verona and Sand Beach that afternoon and survive the ravages of the fire. For example, George W. Jenks was driving a horse and buggy from Bad Axe to Sand Beach. It grew so dark that he could no longer see his horse nor the road. He simply let the horse find the way home, which it did. Also Sgt. Bailey's report to the U.S. War Department stated that news of the conflagration was brought to Verona about two o'clock in the afternoon "by persons who drove almost literally through the flames to reach their homes." He said Verona burned around 3 P.M.

The *Port Huron Daily Times*, interviewed John Ballentine, a Verona businessman, who told how the fire came through his community:

> About 1 o'clock on the afternoon of that day a wind came up from the west that was almost a hurricane, and in an instant the air seemed to be filled with fire.
>
> Mr. Ballentine owned a saw mill, grist mill, store and other property, the whole valued at $50,000. Great care had been taken to protect this property from fire, and around the mill everything was wet down with hose twice a day. But all precautions availed nothing. The fire leaped from one point to another, and in its coming, was so sudden and irresistable that no one thought of saving property. . . .
>
> Mr. Ballentine and his family ran to a cornfield and suffered only from the intense heat and smoke, but all of their property that would burn was destroyed. One peculiarity of the fire was the fact that it skipped some buildings while taking others immediately adjoining.
>
> At Verona, for instance, the parsonage was burned, while the church, but a few feet distant, remained uninjured; and the hotel, the largest building in the place, was also passed over. The church and hotel are the only buildings left there.
>
> On Tuesday scores of people came into Verona, destitute, half naked, and in some cases of children, entirely naked, but there was nothing left there to cloth them with, and scarcely anything to eat.

At Sand Beach

The people of Sand Beach, many of them remembering the terrible fire of 1871, became alarmed around 1 P.M. when the sky turned black and the smoke became stifling. They dug holes and began burying their clothing and furniture.

The *Huron Times*, the Sand Beach newspaper, described that terrible day:

> About 1 o'clock a strange darkness as black as that of midnight had settled down upon this devoted village. Indeed, without any exaggeration, it was absolutely impossible to see one's hand before his face. We have never seen the darkness of a night which exceeded it.
>
> The wind began to freshen from the west and smoke and ashes began to pour in in blinding clouds.
>
> About 2 o'clock the alarm of fire was heard and the fire company were on hand immediately and an incipient blaze in Jenks and Co's salt block was extinguished.
>
> Lanterns began to appear upon the streets and our citizens, who seemed at first bewildered by the sudden change from daylight to total darkness, began to realize that something must be done. All the hose at their command was brought into use and the streets and buildings within reach of the hydrants were saturated with water. [The community had a waterworks and system of water and hydrants throughout the business portion. The use of this water system probably saved the town.]
>
> Water was drawn in barrels to protect the houses upon the outskirts and every precaution taken to guard against a general conflagration in the village and stay the approach of the flames, the dull roar of which could be heard with startling distinctness.
>
> This state of affairs lasted until about 4 o'clock. The darkness in the meanwhile grew more intense and the heat more stiffling. Large cinders and coals of fire now began to fall upon the outskirts of the town and the residents of the suburbs brought their children and women into the center of the town for safety and returned to protect their homes.
>
> About half past four a tremendous roar like a heavy discharge of artillery was heard to the north and

west of the village and the glow of miles and miles of burning material made the western sky red with heat, but failed to cast a ray of light through the thick atmosphere which was blinding and suffocating everybody.

In the meantime the fires were coming nearer with the speed of a race horse and many had given up entirely. They began packing household goods and valuables preparatory to moving them to the lake or burying them. Just as it seemed the critical moment had arrived, when the roar of what was afterwards found to be a terrible cyclone to the west of us had reached its height, a welcome breeze sprung up from off the lake and saved the village from destruction.

The night was a fearful one. Few heads touched pillows and few eyes closed during the long weary hours; and the light of day, which made its appearance after an absence of 18 hours, was hailed with joy.

Mrs. R. C. Allen of Sand Beach, wrote firsthand about the fire in later years.

The sun rose as usual, but very soon it could be seen only as a red ball through a heavy smoke. The fire broke out west of Minden, it became visible there. Before that we did not know where it was, no one could work, because they didn't know what to do.

The barns were full of hay and some of the grain had been harvested and people were frantically fighting that fire for miles away from the shore, as the wind carried it toward the northeast.

People were hiding what they could of things that they might save if the fire came to them, and plowing furrows all around to backfire it, and setting fire on their green stubble which increased the darkness as the green smoldered.

The smoke was so heavy all day we could hardly breathe. It was a long day because no one could content themselves except in fighting fire. Of course, some of the women got out and fought fire.

At Point Aux Barques

As the fire worked its way through Port Austin Township it burned northward to the very tip of the Thumb to the community of Point Aux Barques. From here it traveled east about five miles to Huron City where a lighthouse, known as Michigan Light Station No. 2, was located.

Henry Sill Jr. was the keeper of the light that fall, and his log of September 4, 5, 6, and 7 reflects the anxious hours he and his helpers spent fighting fire.

The log of Sunday, September 4, tells of a fresh southwest wind developing at sunrise and continuing until dusk. Nearly every entry in the log notes the smoke in the air. The lake traffic was quite heavy. Sill records 13 barges, six schooners and 10 steamers having passed on the lake.

On Monday, September 5, the southwest wind returned. Again the atmosphere was smoky. By afternoon, the log shows that the wind became heavy, shifting to the northwest.

Sill recorded that he had his crew at work in the morning painting a rail on the south side of the dock. But in the afternoon, he wrote:

The fire that has been burning for several days in the forests came sweeping down towards the station and seemed to threaten everything in its path with destruction. I had the old surf boat launched and the new one put on the dock ready to launch.

I had everything filled with water, including two pork barrels on the roof of the station, and two buckets ready to use. I had the wagon, oil, tools, spare oars, cart

Devastation caused by fire in Michigan's vast forest areas. Courtesy Michigan History Bureau, Michigan Department of State.

and apparatus taken outside where I thought they proved to be safest from fire. And the powder was put 50 yards from the station down the bank.

Later in the day, Sill determined that the fire no longer threatened the lighthouse, and he began looking around and offering help to his neighbors. He dispatched five men to the Shaw farm, one mile away, but discovered that the farm was destroyed. The fire was burning toward the beach where another family was huddled, and Sill sent a boat to get them out of danger. The family, consisting of a man, wife, five children and a neighbor, were brought to the lighthouse. "As the whole woods and beach is on fire my men cannot go on patrol but will stand their watch at the station," Sill wrote.

The winds of September 6 are recorded as blowing out of the southeast, south and west. For much of the time, the light station was shrouded in heavy smoke from the still burning forests. Sill said that the smoke was so heavy that no ships could be seen passing the station all that day.

> All hands watching fire as the whole country is in a blaze. A 8 P.M. the fire was coming towards the station with such force that I deemed it best to keep every man at station to fight the fire.
>
> We launched the two surf boats . . . and removed everything that we though we could save if the station burned. It was a terrible night. Everything in a flame, We fought the fire all night and saved everything thus far, but the whole country is still on fire.

At last, by Wednesday, September 7, Sill could see through the smoke well enough to begin recording the passing of ships. He put the men to work cleaning up after the long siege, and reported that evening that his "house thoroughly clean and in good repair. Crew putting apparatus back in station as station was not burned."

Paul Geoffrey

As flames roared through Port Austin Township, it approached a farm where Paul Geoffrey, 13, was working in the field with a pair of oxen. The son of a lumberman, Geoffrey's mother had died and he was boarding with people named Niff, who owned the farm.

As the fire approached, the sky grew dark. The trees were being whipped by the hot winds from the southwest. Geoffrey was called from the field by Mr. Niff, who decided the family should get into the well.

Firebrands were falling from the black sky and the trees were exploding into flame as Geoffrey ran across the field toward the well. He saw fire running along a wood rail fence and catch up with a fleeing squirrel.

Geoffrey and four members of the Niff family shared the cramped quarters of the dug, stone-lined well for about two days, while the fire ravaged the forest and buildings overhead. They held blankets and quilts dipped in water over their heads to protect against the fire, smoke and falling debris.

When at last they crawled out of the dank dark well, they had difficulty seeing. Their eyes were damaged from the smoke. The oxen Geoffrey left in the field were dead. All that could be found were the metal parts of the yoke. The buildings were gone. All they had left were the clothes they wore.

Cries of Agony

There were many stories about people escaping the fire by climbing down into wells. The Allen Kennedy family spent that terrible day in their well in Dwight Township, southeast of Port Austin. Rosie Patterson, who still lives in Port Austin, tells the following story as it was told to her:

> Wells were hand dug then. They were quite large and stoned inside like a foundation. . . . The stone was plentiful.
>
> The smoke was so dense, the heat unbearable, but somehow they managed to stay above the water and when timbers fell across the well they would reach out and throw the burning pieces off.
>
> They could hear the cries of humans burning and cattle and horses crying in agony.

George Matthews

When the fire reached the George Matthews farm, 1 1/2 miles west of Redman in Bloomfield Township, Matthews joined his neighbors in a battle for their lives.

In a family history, the story is told how Matthews, his wife Bessie, and their two children, David, 2, and Florence, 5 months, were interrupted at dinnertime when a firebrand fell into the open doorway of their home.

Matthews stepped outside and discovered that the fire surrounded him. It was not long before the house was on fire, and the Matthews family was running down an old logging road, with fire at their heels.

The family reached the Allen McGregor farm and found the house and grain stacks ablaze. Bessie and the children went on to the Robert Scott farm, a little farther down the road, while McGregor and Matthews made a frantic effort to save some of the grain stacks. As they tossed pails of water on the grain, the wind blew a piece of wood striking Matthews in the head. McGregor revived him with a pail of water.

By now the heat of the fire was so intense that efforts by Mrs. McGregor to save some household bedding failed. As she threw the bedding out of the upstairs window of the burning house, it caught fire before it touched the ground. The glass in the house shattered from the heat.

When they realized they were losing their fight, the McGregors joined Matthews in a flight to the Scott farm, where they were joined by Rod Clarin, another neighbor. Here the men took their stand while their families huddled inside the Scott house.

The men fought fire all the rest of the afternoon and through the night and saved the house.

The next morning, half blinded by the smoke and exhausted from the great battle, Matthews, McGregor and Scott began a search of the neighborhood. They found the charred remains of Mrs. Clark, an invalid neighbor, and her son near Willow Creek, at about the place where the Redman School now stands. Five members of the Ripley family were found suffocated in a well.

Ghosts of Forest Bay

Forest Bay was a thriving community on Lake Huron, located midway between Sand Beach and Port Hope, until the fire came. The town was totally destroyed and was never rebuilt. In a sense, it became another casualty of the disaster that swept the Thumb District that fateful day.

Yet the ghost of Forest Bay continues to live. There seems to exist to this day a state of mind among the people in that area of Rubicon Township, many of whom are descendants of the first land owners, of belonging to and being identified with Forest Bay. In fact, Michigan road maps continued to show Forest Bay as a place until the 1950s.

Forest Bay contained a one-room country school, a blacksmith shop, a boarding house that served as a hotel, a store, a dock to receive and send goods, and many homes. The population within a quarter mile of the town was estimated between 100 and 300 people.

The fire came to Forest Bay from the north. Due to the strange effects of two weather fronts that clashed over Huron and Sanilac Counties that afternoon, the line of flame first worked its way to the tip of the Thumb, then circled back on a southerly direction along the coast.

A few miles west of Forest Bay, the two great fire fronts, driven by the two weather fronts, met head-on and caused a roaring sound that was heard for miles. It was the same noise mentioned in the accounts told in Sand Beach, a few miles to the south.

An estimated 15 people died in the forest where this whirling clash of cataclysmic forces converged. Those along the shore simply made a dash for the safety of Lake Huron. They told how they stood in water up to their necks and watched the fire come down on Forest Bay with an unbelievable speed. Long tongues of fire, as though blasted from a blow torch, cut swaths through the woods so quickly that the people thought of saving nothing but themselves. Some families were separated in the confusion. People choked and gagged in the thick smoke.

Thousands of domestic and wild animals shared the lake that day. People all up and down the 60-mile firefront on the Lake Huron shoreline told the same story—how men, bear, deer and domestic livestock coexisted for a few short hours while escaping a common foe.

Lewis Ludington, one of the residents of Forest Bay, was reported killed in the fire. Searchers told Ludington's wife they had found his remains covered with ashes, and his coat nearby in an unburned area.

While grieving the loss of her husband, Mrs. Ludington was visited that evening by a very excited woman who told her that Ludington's naked ghost was approaching. The smoke, haze and light from still burning trees and stumps furnished an eerie setting that was most suitable for ghosts.

On closer inspection, it was discovered that the body in the woods was the carcass of a hog and this strange apparition turned out be a very much alive Lewis Ludington!

The stories that have been handed down do not explain why Ludington and numerous other people were found naked after the fire. Another man, driving a horse and buggy between Bad Axe and Verona the day after the fire, reported finding a naked woman and her two naked daughters on the road. Many accounts told of people walking naked out of the smoldering ashes and into the few towns that escaped the burning.

Did these people tear off their clothes when burning embers set them afire? Or did they act in a moment of insanity, as one man in the Richmondville area reportedly did? It is said that as the fire came upon him, the man kept shouting "Judgment Day" and stripped off all his clothing.

The Alex Nichol Home in Greenleaf Township escaped the 1881 fire. This picture shows the family in front of the home about 1890. Note that there is an absence of trees. Courtesy Lynn Spencer.

Man and Beast

The fire was such a violent threat to the lives of both man and beast that the two often found themselves cooperative partners in a common struggle for survival.

There are accounts of people sharing streams and the safety of Lake Huron with black bears. Other stories indicate that both wild and domestic animals sought companionship with humans when faced with almost certain destruction. It was as if they chose not to die alone.

Forestville, which escaped the ravages of the 1881 fire because of cleared land which surrounded the town, became a sanctuary for livestock. Cows, sheep, hogs and other animals fled from the burning forest and wandered at large in the streets of the town until they were claimed by their owners.

Nettie vanRaaphorst of San Jose, California, told how her mother, a nine-year-old girl living in Forestville, was haunted by the sounds of the displaced animals after the fire. "She often spoke of their mournful and fearful cries in the night. One had the feeling of live wolves at the door!"

Mr. and Mrs. Herbert M. Elder, who live on the same farm on East Mills Road in Marion Township that their family occupied during the fire, tell how the barn escaped destruction. After the fire, all types of livestock, from all over the neighborhood, were found crowded into the basement stable in that barn.

John B. Hall, a Buel Township farmer, stood by his straw stack all night, throwing water on the stack to keep it from burning. The next morning he found a large black bear sleeping beside the barrel of water.

In Elk Township, the Matthew Devitt family battled for hours to save their log cabin from burning. His barns and hencoop were destroyed. When looking over the ruins the next day Devitt found a dead bear beside a log a few yards from the cabin. A pig was found with rabbits and squirrels in the lean-to beside the cabin.

The Aftermath

Charred Wasteland

The fires continued to burn in isolated areas and the charred stumps and logs continued to smolder all day Tuesday and Tuesday night, while survivors carefully ventured out on the charred wasteland. A rain came on Wednesday, September 7.

Now began a time of severe hardship for the survivors. They faced the terrible job of finding their dead neighbors and loved ones and burying them. Those still alive but burned needed medical attention which was almost impossible to obtain.

The very business of survival was difficult. Most of the food was destroyed, and the water in the wells, streams and Lake Huron was covered with ash, making it unfit to drink.

Trees were destroyed, so there was little material left with which to rebuild. Within a few short weeks, the first hard frost and then the snowfall of winter would be arriving. Not only did these people have no home, many had no shoes or coats. They possessed only the rags they wore on their backs.

The air soon became polluted with the smell of dead and decaying animals. It would be months before this smell went away.

Worst Fears Realized

The first news reports on September 6 were fragmented, but it is clear that the writers knew that something terrible had happened in the Thumb.

The *Detroit Evening News* told of "terrible fires raging in the forests northwest and north of Bay City, . . . air full of cinders, . . . people suffering from heat and smoke." The *Detroit Free Press* said "the Saginaw Valley for miles is a sea of flames."

In Port Huron, the stories were starting to trickle in and the September 6 edition of the *Port Huron Daily Times* gave a more dramatic account:

> The indications are that the worst fears regarding the results of the continued drouth to the country lying north and west of Port Huron have already been realized, or will be within the next day or two. Both the telegraph lines extending into Huron and Sanilac Counties are down, and nothing definite can be learned, as all communication with those sections is cut off.
>
> The P. H. and N. W. train northward Monday morning could not get farther than Deckerville, and in returning was stopped at Carsonville by the fire. Number 4 left Monday afternoon and was compelled to stop at Croswell, which is the furthest point northward to which communication is now open.
>
> The propeller "City of Concord," Capt. Frank Hebner, left Port Hope shortly after 3 o'clock on Monday afternoon. At that hour, Capt. Hebner says the town appeared to be nearly all on fire. The dock was burning

and he was compelled to leave it. A number of people also left the town by his boat. The air was dense with smoke and it was so dark that it was necessary to light the lamps.

> From Port Hope to Forester the shore seemed to be one continued blaze. Sand Beach and Port Austin were also suffering from the fire. Captain Hebner was almost blinded with the smoke and heat, from the effects of which he had not recovered this morning. When he left Port Hope he had the schooner "L. L. Lamb" in tow, but during the darkness she got fast in a floating raft and could not be got out. The propeller was also entangled in the raft for two hours.
>
> Persons who have traveled through the central and western portions of St. Clair County during the past two or three days report the woods still on fire in every direction, the country covered with a pall of smoke, and heavy losses of property.
>
> A dispatch from Forester, Sept. 5, says: "Forest fires are ranging in every direction. A strong west wind blowing all day has brought the fire in dangerous proximity to this town." It was only by the most persistent efforts that this town has been saved and the danger is not yet over. The smoke is terrible. About noon it began to mount, and by 3 o'clock, total darkness prevailed, lasting till about 5 P.M. when it cleared up slightly. Lights have been burning constantly since 2 P.M.
>
> Great destruction of property is reported from all quarters, in many instances accompanied by loss of life. it is reported that fires have made a clean sweep of the town of Deckerville, near Black River. Probably 75 or 100 families in this part of the country are left homeless tonight.

"Burnt till they was black"

Elizabeth Gifford, wife of John Gifford from Sand Beach, wrote to her mother on Thursday, September 8. In her letter, she gives a vivid account of the horrors the fire left in its wake:

Cass City was one of the few spots spared by the Great Fire of 1881. The upper left-hand picture shows volunteers in Cass City distributing clothing to fire victims of the area. Courtesy American Red Cross.

We have had a horrible fire here from people clearing their lands, and also a tornado. Last Monday at one o'clock in the afternoon it got as dark as any night ever was known and we light our lamps and kept them lit till the next day. Could not see our hands before our faces outside. The fire burned Verona, Bad Axe, Caro, Cass City, Huron City, Forest Bay, Port Hope, Richmondville and mostly all up only left a very few houses standing. Burnt all of Sanlake County, Tuscola County and hundreds of people.

They brought 12 (*bodies*) here burnt till they was black. John helped bury them. They are fetching loads of people from the country all burnt alive. And hundreds of cattle, sheep and hogs lay dead along the road. Some with their legs off and some going along with their feet off and living yet. It burnt the telegraph and blocked the (*railroad?*) cars for two days. . . .

I never saw better crops before and the fire burnt all of them. And also our place. Hundreds of acres almost ready for the plough.

They buried 18 all in one grave in Paris and some families never have been heard of. Some jumped in wells and was saved and some was killed. We was packed up and had our things ready to bury them. There was boats along the shore waiting for us. We was afraid of getting swamped—with the fire as the wind looked a little swift—and blew it back.

Some little children in this town has got their feet burnt very bad. It is the saddist sight ever I heard of. Excuse the scribbling for I am very nervous for fear of getting burnt yet.

"I can't see you; I'm blind!"

Elder John J. Cornish, a priest for the Reorganized Church of Jesus Christ of the Latter Day Saints, in his autobiography, graphically described traveling north from Lakeport to his home at Richmondville on the day of the fire.

Cornish explains in his book that he and another church member bought a small sawmill which he hoped would support his church ministry.

"There was considerable timber left that had not been picked up by the big lumber companies. The land was now being taken up by farmers, all of whom would need lumber for their buildings, fencing, etc," Cornish wrote.

On the day of the fire, Cornish began a trip by horse and buggy from Lakeport north to Richmondville. The rig was being driven by a friend, Lyman Whitford.

"We had not traveled far when Bro. Whitford said 'I'm afraid there is a big fire farther up the shore; it is so smoky the last few days. It looks very much as it did ten years ago when we had the terrible fire.'

"We traveled on. By and by we met a man who lived about 20 miles farther north, who informed us that the whole country north of us, both Sanilac and Huron Counties, was on fire, and that some people and much livestock had been burned."

By now, Cornish was concerned about the safety of his wife and three children, especially when another traveler told them that Richmondville was in ashes.

Whitford and Cornish traveled on until they came to places where flames still crackled along the sides of the road, and then into the burned area where logs and stumps still burned.

The fire, coming as it did, gave people and stock an opportunity to stand in the open when those places were burning, and when the worst was over, and the fire came through those streaks not already burned, the people and stock could go back on the burned places.

About 9 o'clock in the evening we reached the place where the village of Richmondville once was. I learned that my dear wife and children were alive . . . but that all of our property was destroyed.

Out of the 27 houses from the dock on the road, running west to where the land was not yet taken up by settlers, stood only three buildings that were not burned.

As he walked among the survivors searching for his wife and friends, he heard: "Cornish, I know your voice, but I can't see you. I'm blind." And, "Well, elder, I was well off two days ago. I am not worth a dollar today." And, "Give me your hand, Johnnie. Your mill and house, and all my buildings and fencing are in ashes."

Still searching for his family, Cornish, with Whitford, drove west one mile past the ruins of the mill until they found a small log house that survived the fire. There they found the Cornish family, huddling with many other fire victims.

"By the dim light of evening we could see a family here, another huddled there, together. Some were bareheaded, others barefooted, some with barely enough rags left on them to cover their bodies. . . . It was among this bunch I found my wife, weeping, with the little baby in her arms. We greeted each other with a feeling that I can not express; with sorrow for our losses, yet glad and overcome with joy that our lives were spared to meet again."

As he stood there and talked with his neighbors about the fire, Cornish said his little boy tugged at his pant leg and said "Papa, I am hungry, and I want to go home."

"Poor boy; we had no home, and nothing to eat. How should we live when everything was burned? All wooden pails were burned; all pails and dishes were unsoldered and broken up. But we had some iron kettles and some other vessels that would hold water. These we used for milk pails (as some cows were still alive) and to boil potatoes. One kettle would answer for two or three families."

The Doctors

While many people treated themselves or were helped by neighbors, the few doctors in the Thumb were on the job.

The physicians were A. M. Johnson and Charles Davis at Sand Beach, Ryerson G. Healy and A. N. Johnson at Minden, A. T. Kay at Tyre, W. P. Brown at Lexington and Alderton in Deckerville. There undoubtedly were others whose names escaped the records of that day.

Stories were told how these doctors worked around the clock treating the wounded and searching the still smoldering area for fire victims who might still be alive.

One of the Sand Beach doctors was said to have placed 11 charred bodies on his own barn floor while he continued working to save the living.

Several doctors were sent from Port Huron to help in the work after the first stories of the fire reached that city. Homes and public buildings that escaped the fire were turned into temporary hospitals and hotels.

"Help Us!"

The area newspapers that managed to publish that first week gave graphic descriptions of the horrors the fire produced. The papers also issued cries for assistance.

A Relief Party Searching for Sufferers in the Wake of the Fire. A sketch based on drawings from corresponding artists for Harpers Weekly, *October 1881. Courtesy American Red Cross.*

The *Huron Times* in Sand Beach carried a headline "Come Over and Help Us." The story said:

> We are in sore distress. Our farmers are penniless. They have neither seed, clothing, nor that upon which to subsist through the coming winter. They have no buildings, fences nor lumber to construct them. Their cattle and implements are destroyed. Their money, if they had it, was licked up in the general conflagration.
>
> Here is a work for the charitably inclined and benevolent people all over the country. In sending supplies to this county you are doing a Christian duty. Supplies can be sent to J. Jenks and Co. of this place who will see that they are properly turned over the relief committees for distribution. Some arrangements will be made for freight. Our merchants have opened their stores and sent supplies with an unstinted hand, but their whole stocks are inadequate to the demand.

At Minden, the north Sanilac County town that miraculously escaped the fire while everything around it burned, a relief committee was organized on September 6, the day after the fire. The following day, George McDonald read a letter of appeal from the Minden committee to a Port Huron relief committee which was holding its organization meeting:

> Thousands are destitute of clothing and suffering from hunger. Wheat elevators containing the marketed crops are consumed. The village of Minden escaped by a miracle and hundreds are finding their way here in a starved condition. A public meeting was held here tonight for the purpose of taking steps to relieve the sufferers, but our limited means force us to call on a generous public for assistance. Mr. George McDonald, a worthy citizen, was appointed to convey this report of the undersigned (who were delegated at said public meeting a committee to prepare this petition) to Detroit and other places to solicit aid which is necessary to keep persons from starving. Over 200 persons are burned alive and a large amount of livestock. Send all relief, either money or goods, to George McDonald, Minden, Sanilac County.

The letter was signed by William A. Mills and J. H. Shults.

Another letter from a Sandusky business, Corbishly and Doyle, written September 7 to F. Saunders and Co., Port Huron, was published the next day in the *Daily Times*:

> We write to you to say that the forest fires have destroyed our county. It is a black, burned waste. Every settler in Watertown, Elmer, Moore and Custer, with hardly an exception, has lost his all; houses, barns, grain, cattle, and what cattle are left will have to be sent out of the county, as there is no feed.
>
> Five dead bodies were picked up in the road yesterday in Watertown, and 18 more in Moore and Argyle. The suffering and distress is terrible. We ask your aid and assistance for our unfortunate people.
>
> You can publish this in the *Port Huron Times*, and, if no general system of relief is organized, we would like to have you do what you can. We will distribute anything you may send in the line of clothing, bedding, second hand cook stoves, meal, etc. Act at once.

John V. Shepherd, who was burned out of his farm one-half mile south of Ubly, wrote on September 7 to a friend M. H. Allard of Port Huron:

> The country is swept in certain directions almost clean of everything. Many are left with scarcely enough clothing on them to hide their nakedness. We are burnt somewhat, some badly, and some are sick. We need medical aid as well as provisions. Will you please make our case known to people of influence in Port Huron?
>
> Mills, stores, bridges and almost all farm buildings

and fences are burnt as far as we have any knowledge. Tom rode up to Verona Mills yesterday. Almost the entire village and vicinity are gone and it is feared that north of there, on Hubbard's low lands, things are even worse. My two sons-in-law there are swept clear of all but a few head of cattle.

The *Detroit Post and Tribune* received a dispatch from a correspondent in Tuscola County on Thursday, September 8 which read:

> The situation in Tuscola and Huron Counties is growing more desperate. Near Cass City a section eight miles square was burned over and nine lives were lost. The people are suffering terribly. Mayor Welch has called on the people here for aid and an organized effort will be started tomorrow.
>
> Heavy rains are reported north and west. The smoke is dense and more disagreeable than on any former day. Nothing has been heard from the Indian settlement on the Cheboyganing. It was totally surrounded by fire yesterday.
>
> Near Vassar, Tuscola County, the fires have abated but the losses in that section have been heavy. In Saginaw and Bay counties the fires have also abated.

A second story in the *Detroit Post and Tribune* that same day told of conditions at Port Sanilac:

> A large party of men left here this morning to bury the dead beasts in Forester Township. Dr. Hoyt returned after more medicine and reported several badly burned and many more so blind that they had to be led. Many are without food. Numbers of parties left here today with food, clothing, shoes and medicine. More help must come quick, or much suffering will ensue.
>
> George Furguson of White Rock, who has been on the road since Monday, reports he has seen 116 burned bodies. At one place he saw four wagons bearing eight coffins, one man walking behind alone, it being his entire family. Another man was following three coffins.

The Port Huron and Northwestern Railway Co. offered free shipments of relief supplies into the burned area. The company also offered benefit rides from Port Huron to Sand Beach, to enable citizens to look at the devastation. The fare of two dollars was used for the relief fund.

Port Huron Relief Group

Port Huron's Mayor E. C. Carleton called a meeting of citizens on Wednesday, September 7, to discuss assistance to the burn victims.

The town's important citizens and leading businessmen attended the meeting, which turned into an evening of speech making. The group received an emotional charge when George McDonald of Minden read an appeal from his community for help.

A resolution was adopted creating a relief committee, with Mayor Carleton as chairman. The committee was charged with soliciting for food, clothing and money.

Carleton appointed 18 members, three from each of the city's six wards, plus a six-member executive committee. The committee immediately drafted a public appeal, which was published in the *Port Huron Daily Times* the following day. The letter probably was sent to newspapers all over Michigan and surrounding states. It read:

> To the People of the United States: A most appalling disaster has fallen upon a large portion of the counties of

Huron and Sanilac, Michigan, with some adjacent territory, a section of country recently covered with forests and occupied by nearly fifty thousand people, largely recent settlers, and either poor or in very moderate circumstances.

In the whole of this section there has been but little rain during the past two months and everything was parched and dry, when on Monday, September 5, a hurricane swept over it, carrying with it a sheet of flame that hardly anything could withstand. We have reports already of over 200 persons burned to death, many of them by the roadside or in the fields while seeking places of safety, and it is probable that twice this number have perished.

We also have reports from 20 or more townships in which scarcely a house, barn or supplies of any kind are left, and thousands of people are destitute and helpless. All of these people require immediate assistance, and most of them must depend upon charity for months to come.

We are doing all in our power to succor them, but the necessities of the case are so great that the contributions of the charitable throughout the country will be required to help them through the winter.

We therefore appeal to you to send money, clothing, bedding, provisions, or any other supplies that will help maintain the sufferers and enable them to provide shelter for themselves and begin work again on their farms.

Contributions may be sent to the chairman, secretary or treasurer of the relief committee appointed by the citizens of Port Huron, whose names are signed hereto, who have sent agents through the burned district to ascertain the wants of the sufferers and distribute supplies.

Among the signers were Carleton, M. H. Allardt, secretary, and H. G. Barnum, treasurer. Another member was O. D. Conger, U.S. Senator.

Drawings like this began appearing in newspapers and magazines of the day when publishers learned of the plight of the Thumb pioneers who had been scourged by the Great Fire of 1881. The appeals for aid brought money and clothing from all over the nation. This is from Frank Leslie's Illustrated Newspaper, *October 1881. Courtesy American Red Cross.*

The committee issued a second appeal on September 11, which was more emotional and undoubtedly was written by people who visited the burn area:

> To the American People: We have tonight returned from the burnt district of Huron and Sanilac counties. We have seen the burned, disfigured and writhing bodies of men, women and children. Rough board coffins containing the dead are followed to the grave by a few blinded, despairing relatives; crowds of half starved people at some of the stations are asking for bread for their family and neighbors.
>
> We heard of more than 200 victims already buried and more charred and bloated bodies daily discovered. Already more than 1,500 families are found to be utterly destitute and homeless. They huddle in barns, in school houses, in their neighbor's houses, scorched, blinded, hopeless.
>
> Some still wander half crazed around the ruins of their habitations, vainly seeking their dead. Some in speechless agony are wringing their hands and refuse to be comforted. More than 10,000 people who, only one week ago, occupied happy, comfortable homes, are today houseless, homeless sufferers.
>
> They are hungry and almost naked when found, and in such great numbers and so widely scattered that our best effort and greatest resources fail to supply their immediate wants. Without speedy aid many will perish and many more will suffer and become exiles. Our people will do their utmost for their relief, but all our resources would fail to meet their necessities. We appeal to the charity and generosity of the American people. Send help without delay.

The *Port Huron Daily Times* reported on September 9 that the relief committee had received about $2,000 in cash donations, and had already shipped 17 barrels of flour, 350 loaves of bread, 593 shoulders of beef, 127 hams, three barrels of pork and 42 boxes of clothing and general merchandise by rail.

On September 13, another shipment of supplies went north from Port Huron. It included 8 boxes of clothing, 4 barrels of pork, 80 sacks of meal, 2,500 loaves of bread, two gallons of whiskey for the hospitals, one dozen brooms, three cases of medicine, two gallons of olive oil, two gallons of carbolic acid, four gallons of lime water, one keg of lime and 25 barrels of flour.

Similar lists were published in the weeks that followed. Help was not long in coming to the disaster area.

The Rivalry

The national appeal for help to the burned district was overshadowed that month by the shooting and eventual death of President James A. Garfield on September 19. Nevertheless, the disaster was big enough that it did gain the attention of the nation.

Relief Committees similar to the Port Huron group were organized in Detroit, Saginaw, East Saginaw, Bay City, Flint, Sarnia and numerous other area cities. Money and shipments of food and supplies poured into the area from all over the United States.

Stories about the fire were appearing in all of the major newspapers including two popular magazines of the day, *Harper's Weekly* and *Frank Leslie's Newspaper*

Problems in receiving and distributing food and supplies were of immediate concern. Every committee was sending representatives into the burn district to determine the needs and secure agents and terminal points. Some communities in the burn area quickly appointed agents to receive the assistance as it came to them. But rivalries were developing.

The *Port Huron Daily Times* reported on September 17 that J. S. MacDonald, a representative of a New York City relief effort, was in Port Huron pledging cash assistance. The *Times* said MacDonald "thought it most strange and unfortunate that Detroit should have sent out a call for all money and supplies to be sent there, and that the appearance of rivalry thus raised would injure the cause."

The article was referring to a proclamation by Michigan Governor David H. Jerome on September 15 that all money and supplies be sent to Detroit Mayor William G. Thompson. In the meantime, Jerome toured the disaster area listening to complaints from the Port Huron Committee. On September 19, he issued a second proclamation, crediting the Port Huron Committee with being on the scene first and doing an effective and efficient job of helping the distressed.

Heated editorial barbs were appearing daily in the *Port Huron Daily Times* and the Detroit newspapers. The Port Huron Committee called it "a disgraceful wrangle over a charity."

In a report published the following year, the Port Huron Relief Committee expressed continued bitterness for what had gone on between the two cities:

> While Detroit justly prides itself on being the metropolis of the state, it has, nevertheless, always looked with fear upon its small neighbor, Port Huron. The Port Huron and Northwestern Railway, built by Port Huron energy and capital, had been completed through the center of the burned district, drawing much valuable trade from Detroit.
>
> Politically, too, Port Huron had just scored a point against Detroit, by the election of Omar D. Conger to the United States Senate. These facts intensified the business and political rivalry already existing between the two cities. Add to this, that the relief matters were assuming greater proportions than could have been anticipated when they were inaugurated, naturally caused the boiling pot to bubble over.

On September 22, the Detroit Committee asked for a meeting with representatives of the other Michigan organizations in an attempt to resolve the problem. But the Port Huron Committee boycotted the meeting.

On September 29, when the Detroit Committee resigned, the Detroit newspapers appealed to the governor for a merger and creation of one state relief organization.

A conference was held in Detroit on October 5, and three representatives of the Port Huron Committee attended. The delegates agreed to merge the work of the Detroit, Port Huron, Saginaw and Bay City committees and that all supplies be distributed under the direction of the Port Huron organization.

Governor Jerome issued a proclamation on October 6 creating the Michigan Fire Relief Commission. He named Henry P. Baldwin, a Detroit banker and former governor, committee chairman. Other members were Alexander H. Dey and D. C. Whitwood, also Detroit bankers; George C. Codd, Detroit postmaster; F. W. Swift, a Detroit businessman; U. S. Senator Omar D. Conger from Port Huron, and Charles T. Gorham, a Marshall banker.

The Port Huron Committee immediately cried foul and

withdrew from the state organization. Its 1882 report complained that

> the extreme unfairness of the composition of the (*state*) commission is apparent. Detroit received five members, Marshall one, (who has never taken an active part with the commission) and Port Huron one, in the person of Senator Conger, who was then in Washington, with the certainty that his duties would keep him at the Capitol during the winter.
>
> Notwithstanding the evident intent of Gov. Jerome to slight the Port Huron Committee, there was no disposition to break the compact made at the conference until Messrs. Baldwin and Dey appeared at Port Huron and demanded the unconditional surrender of all moneys in the hands of the Port Huron Committee, entirely ignoring the third resolution passed at the Conference.
>
> The demand was decidedly at variance with the wording and spirit of the resolutions. After a protracted meeting, and after various propostions for a compromise had been made by Port Huron, (Mr. Baldwin still insisting upon a surrender of the money before he would make any concessions), the Port Huron Committee broke off negotiations, and resolved to wind up its own affairs.

Thus, two separate committees worked through the winter and most of 1882, providing at first food and clothing, and then lumber and building materials and finally seed for crops. The Michigan Legislature appropriated $260,000, plus another $15,000 for construction of 46 schools in Sanilac, Huron and Tuscola Counties. The schools were all one- or two-room buildings, costing between $500 and $600 to build.

For distribution purposes, the state committee divided the burned area into 20 districts, with agents assigned to each. The Port Huron Committee organized 18 districts and assigned agents to each. The districts were mostly alike, and in many cases, the same agents served both committees.

The centers were located at Bad Axe, Cass City, Caro, Carsonville, Croswell, Deckerville, Forester, Forestville, Marlette, Melvin, Minden, Paris, Port Austin, Port Hope, Port Sanilac, Sand Beach, Vassar, Verona, White Rock and Valley Center.

In spite of the problems, and perhaps stimulated by them, the two committees successfully handled combined cash donations of $1,012,648 and contributions of goods estimated at about $500,000 by the end of 1882.

But the rivalry that developed between Port Huron and Detroit pointed to a need for some kind of national organization capable of supervising relief efforts during times of major disaster.

Such an organization was being organized that same year in Dansville, New York: The American Red Cross.

Clara Barton

Clara Barton carved an indelible mark on American and world history. A woman with great compassion for suffering people, she had a driving will to do something for them.

Barton was active in organizing volunteer aid for Civil War battle wounded and personally went into the field to minister to stricken soldiers, both Union and Confederate. She later went to Europe where she worked as a representative for the newly organized International Red Cross Committee during the Franco-Prussian War. There, she admired the Committee's organized efficiency, in contrast to the unpreparedness and delay she had seen during the American Civil War.

MICHIGAN'S
TERRIBLE CALAMITY.

DANSVILLE SOCIETY OF THE

✚

RED CROSS.

A CRY FOR HELP!

The Dansville Society of the Red Cross, whose duty it is to accumulate funds and material, to provide nurses and assistants if may be, and hold these for use or service in case of war, or other national calamity—has heard the cry for help from Michigan. Senator O. D. Conger wrote on the 9th of September that he had just returned from the burnt region. Bodies of more than 200 persons had already been buried, and more than 1500 families had been burned out of everything. That was in only twenty townships in two counties. He invoked the aid of all our people. The character and extent of the calamity cannot be described in words. The manifold horrors of the fire were multiplied by fearful tornadoes, which cut off retreat in every direction. In some places whole families have been found reduced to an undistinguishable heap of wasted and blackened blocks of flesh, where they fell together overwhelmed by the rushing flames. For the dead, alas! there is nothing but burial. For the thousands who survive, without shelter, without clothing, without food, whose every vestige of a once happy home has been swept away, haply much, everything, can be done. The Society of the Red Cross of Dansville proposes to exercise its functions in this emergency, and to see to it that sympathy, money, clothing, bedding, everything wh'c'i those entirely destitute can need, shall find its way promptly to them. But the society is in its infancy here. It has in fact barely completed its organization. It has not in possession for imme liate use the funds and stores which will in future be accumulated for such emergencies. It calls therefore upon the generous people of Dansville and vicinity to make at once such contributions, money or clothing, as their liberal hearts and the terrible exigency must prompt them to make. Our citizens will be called upon for cash subscriptions, or such subscriptions may be left with James Faulkner, Jr., Treasurer of the Society, at the First National Bank of Dansville. Contributions of Clothing and Bedding may be left at 154 Main street, Maxwell Block, Sewing Machine Agency of Mrs. John Sheppard.

☞A special agent of the Society will be dispatched with the money and goods to see to their proper distribution. Please act promptly.

EXECUTIVE COMMITTEE RED CROSS.

Dansville, Sept. 13, 1881.

DANSVILLE ADVERTISER STEAM PRINT.

On September 13, the nascent American Red Cross chapter formed by Clara Barton of Dansville, N.Y. sent this plea for help in the local newspapers. Courtesy American Red Cross.

When she returned to the United States in 1873, Barton was in poor health. She made her home near Dansville, New York, to take advantage of the Jackson and Austin Sanitarium located there.

It was not long before Clara Barton was working hard to found an American Red Cross. She succeeded in getting President Garfield and James G. Blaine, U. S. Secretary of State, interested in such an organization. And she had one other important supporter: Senator Omar D. Conger of Port Huron.

Conger was a congressman when he first met Clara Barton in the fall of 1877. She was active that year in promoting the American Red Cross Committee and no doubt crossed paths with Conger during one of her trips to Washington, D. C. Blanche Colton Williams, in a biographical account of Barton's life, said a lasting friendship began when Conger discovered that Barton had nursed his brother at Fredericksburg during the Civil War.

There were two unsuccessful attempt to organize the Red Cross on May 12 and again on May 21, 1881, in Washington, D. C. Both meetings were in Washington, the first was held at the home of Senator Conger, the second at Barton's residence. Work was temporarily halted, however, when President Garfield was wounded by an assassin's bullet in July, and until his death on September 19, 1881.

Conger, however, remained a loyal supporter of the Red Cross concept. On May 17 he introduced a Senate resolution directing the Secretary of State to prepare translated copies of the Articles of the International Red Cross Convention signed at Geneva, Switzerland, August 22, 1864, along with forms of ratification developed by several governments. Conger's resolution was passed unanimously and led to the ratification of the Geneva Convention by the United States the following year.

The Geneva Convention, and its subsequent ratification by the nations of the world, has served to this day as the cornerstone for the work of the International Red Cross among wounded and sick soldiers and prisoners of war.

Barton returned to Dansville in the summer of 1881, still determined to accomplish her dream of organizing an American Red Cross.

On August 22, only 14 days before the Michigan disaster, she succeeded in organizing a local chapter of the Red Cross in Dansville. It was purely a local effort, however, because the Red Cross was not recognized by the federal government and would not be made part of the international organization for another seven months.

Senator Conger spent the summer of 1881 in Port Huron and was there when the fire swept across the Thumb District. Mayor Carleton named Conger a member of the Port Huron Relief Committee.

It is not known whether Conger and the relief committee made an appeal directly to Barton, or whether she read of the disaster and decided to offer unsolicited help. Whatever the reason, letters were exchanged between Barton and Conger, and she sent money directly to Conger to be used in the relief work.

While the International Red Cross was organized to assist only in times of war, Barton believed the organization had a purpose in peacetime as well. She saw the Red Cross as a vehicle for getting relief to people during times of major disaster, whether natural or man-made.

C. H. Whelden Jr., director of research information for the Red Cross in 1955, wrote that "the related contents of several letters suggests that Miss Barton saw a possible opportunity in the Michigan forest fires disaster to attract public attention to the newly organized Red Cross association." However, Whelden pointed out ". . . the organized work of the local relief committees offered no appropriate opening for effective participation by the National Red Cross." But the children of the Peter Sparling family at Tyre have passed down the following story about the days following the fire: "We prayed constantly for food and clothing to see us through the winter, as it meant survival. And when the word came that a ship loaded with food and clothing arrived at Forestville, a few miles from our home . . . it was just like the heavens opened up! The only reason we are here today is due to help from the Red Cross."

The story, as told by Myrtle Sparling Brunk to other family members, credits the Red Cross for a boat load of food it probably did not send! The boat may have been the revenue steamer "Perry," which was sent by the Detroit Relief Committee into the burned district with supplies.

Many stories, like Myrtle Brunk's, crediting the Red Cross for providing assistance, were reported, given the fact that the Red Cross has grown to be an organization synonymous with disaster relief. The residents of the Thumb district may not have known the exact source of their help in those terrible days, and assumed later that it was the American Red Cross.

While it did not play a major role in the local relief work, Barton's Red Cross did remarkably well for a small organization as far away as the Dansville Chapter.

A *Daily Times* story on September 17 reports Barton's interest in the Michigan fire: "Senator Conger received a letter this morning from Miss Clara Barton, the American representative of the Red Cross relief organization of Geneva. She has been in Rochester, N. Y., and other places stirring up the people to a sense of their duty and urging them to contribute largely, and has met with good success."

Barton may have used the Michigan disaster as a vehicle for generating more interest in the Red Cross work in New York State, for it was not long before second and third chapters of the Red Cross were formed at Rochester and Syracuse. All three chapters worked to raise money and supplies for the burned district.

Barton contacted Julian B. Hubbell, a medical student at The University of Michigan, requesting that Hubbell become the Red Cross field agent in the burned district.

Hubbell and Mark J. Bunnell, of Dansville, visited the Thumb area. Bunnell probably brought the first invoice of money to Port Huron when he arrived on October 4. The Red Cross eventually raised about $80,000 in money and supplies nearly all of which was turned over to the Port Huron Relief Committee.

If Clara Barton saw the fire as a means of putting her Red Cross to work in the field, she must have been disappointed when Hubbell and Bunnell discovered their services were not needed.

Hubbell wrote from Sand Beach on Oct. 6 that he had "been to Port Hope ten miles above here. Find relief well organized wherever I have been. Greatest need is for beds, bedding, building. Very little sickness. Most of those who were injured are recovered."

On the same day, writing on the letterhead of the Port Huron Executive Committee for Relief, Hubbell wrote that he found "a well organized system for relief at work

here. . . . Maj. Bunnell has come in from the burnt district. He is well pleased with the manner of work and says that it is so well organized that it would not be well to interfere. The people see the good that would come out of a permanent organization similar to what they now have, and I think would gladly accept the Red Cross when they are relieved from the present pressure of work."

Three days later, while at Minden, Hubbell wrote: "I have been trying to plan some way by which the Red Cross could work promptly and still not collide with the present relief."

Then on October 11, Hubbell wrote from Detroit: "Called on Port Huron Committee yesterday. They have large supplies of clothing on hand and are now sending the goods to the front. They think there is no need of any more being sent. They can do more good with money than any other contribution now. . . . I see nothing here now that the Red Cross can now distinguish itself in."

Digging in for Winter

Julian Hubbell painted a graphic picture of conditions in the burned district one month after the fire when he wrote the following report from Minden as a field agent for the Red Cross. The letter was dated October 9:

> The Michigan sufferers have truly been tried by fire. One can tell others but little that he here sees and learns of the suffering that has been and still is endured; for it cannot be believed until one sees for himself, examines the extent of the burning, the amount of combustible material, etc.
>
> The ground was so dry that one and a half million acres [a slightly exaggerated figure] of it were burned

over in five hours. The air became so hot that men and women, cattle and horses, burned to death in open fields twenty and even fourty rods from where the fire reached. The day became dark as blackest night and so thick with smoke that even the fire could only be seen for the distance of a few feet. Burning cinders fell as from the heavens, and blew over the ground like snow.

Friday and Saturday I spent in visiting the country about Port Hope, Sand Beach, and Minden. The fields are green with wheat growing, in some cases, upon new ground cleared since the fire. Some of the fields are already fenced. While the inhabitants have been doing this work, they have lived in root cellars six by ten feet; in rail pens without cover, or furniture or bedding, and sometimes with one or two comforters for a family of eight or ten persons.

Since the seed has been gotten into the ground, they have begun to make shelter for the winter. In driving a distance of twelve miles I counted fifty-one new dwellings, built since the fire five weeks ago. An idea of the destruction of dwellings throughout the country may be gathered from the fact that in eight miles I counted sixty-one ruins of dwellings, besides barns.

I found one family of eight living in a milk house, six feet by eight feet, with only straw to sleep on. In another house, fourteen feet by twenty feet, eighteen persons were living, whose only furniture consisted of two chairs, one square table, one bedstead, mattress, four comforters, two spoons, two cups, six plates, and a stove. A house eight feet by thirteen feet and six feet high, with a small government tent, protected thirteen persons. In this they not only lived themselves, but cared for an old lady 97 years of age, a fanning mill outside serving for a pantry. Dough fried in pork grease was their dinner, all cooked in a small frying pan. I saw a woman sitting on the ground sewing.

The families here average six persons. From supplies received, each family has been supplied with goods about as follows: besides provisions, one stove to two families, a fair supply of thin clothing, two blankets or comforters to each family, two knives and forks, two spoons, two cups and saucers, three plates, two chairs. One thousand feet of lumber are now being distributed to each family.

Articles most needed, aside from provisions, are beds, bedding, blankets, ticks or ticking for same, woolen shirts, woolen underwear, heavy boots and shoes and chairs. Horse feed will be much needed. Fence wire to protect crops will be a great help. Farmers can aid much by saving seed corn now for spring planting. Grass seed will also be wanted; early vegetable seeds and garden seeds of all kinds. Let gardeners and farmers think of this.

The only way to realize the situation is to keep constantly in mind the fact that everything that could possibly burn has been swept clean from the ground; very few having anything left but the clothing that was upon them at the time, and as destitute as if transported in a moment to a desert. Not a bud for a rabbit to browse, and not a rabbit to browse if there were.

These people must be supported until next harvest. How much will this require? One thousand barrels of pork or flour would last them twelve days, if each person had one pound a day. From this statement any one can estimate for himself the amount of provision that must be supplied. . . . There is little danger of getting too much of anything except summer clothing.

Surviving

Mrs. Florence Peters of Lincoln Park, who furnished the story of the William White family, also wrote a description of how the Whites lived during the months following the fire. The family lived in Austin Township near Tyre.

The Whites were one of about nine families that moved

into the only house left standing in the area, somewhere near Holbrook. There was little food. The women and children lived in the house and the men stayed in the barn.

> For weeks they stayed there until relief started to come in. My grandfather got enough lumber to build a lean-to shake, but not enough for a door. So they nailed quilts up to it and lived in such a place all winter. Bunk beds at both ends clear to the roof, a plank table and old hi-oven stove in the middle.
>
> When the house burned the stove had fallen down in the cellar and broke one leg off. That was fixed with some bricks. There were some other things down there, cooking utensils and a two gallon crock of blackberry jam with the oiled cloth burned off. They scraped the ashes off and ate the jam which was black and thick like liver.
>
> They were better off than many others after that fire. Relatives in Canada shipped them clothing for the family and some bedding.
>
> The cattle that were turned loose . . . some survived in the river and some had such burned feet they had to be killed. That meat was hung up and dried and ate that winter.
>
> Rags were wound around childrens feet, or pieces of old coats or pants made into a kind of moccasin to keep feet from freezing in a shack with only a dirt floor. The first shoes my mother had for the next winter were a pair of copper-toed ones. She never forgot them. They hurt her feet. They came in relief and she was the only one they fit.

Lasting Scars

The effects of the Great Fire have remained in the Thumb area for 100 years. Stories are still told and some older natives can show charred roots of once stately white pines that have not completely rotted away.

The scars are slow in disappearing. For about 30 years the landscape was almost barren of trees. Early photographs show young saplings in front yards, and acres of treeless fields beyond.

Yet for all of the terrors, there were some good things to remember about the Great Fire. In many areas the great piles of slashings and dead tree trunks, and the thick second growth of forest that followed the 1871 fire, were totally consumed so that the rich and fertile land was opened overnight to the farmer's plow. Within weeks, wheat was growing and turning the blackened wasteland into fields of green vegetation. Land values actually increased.

The land was so fertile, in fact, that many crops could be grown on it. Michigan's Thumb today is known for its navy beans, sugar beets, red wheat, peas, cucumbers, corn and hay crops. The area boasts a prime dairy industry. Huron County is known as the navy bean center of the world. Rich sources of peat moss were found in swampy areas and this rich humus, prized by gardeners, is sold as a cash crop.

Sod farms were developed on the rich lowlands. Canning and food processing plants were built. Towns that began as lumber camps flourished again as agricultural centers.

Ironically, even though the fire burned homes and barns, it saved years of labor in land clearing. The agony, the death and maiming of so many friends, neighbors and animals, and the loss of personal belongings in the homes, were the massive prices paid, however.

A second railroad was soon built in the area and probably because of the fire. The Pontiac, Oxford and Northern, now the Grand Trunk Railroad, was laid north through Imlay City, Cass City and Huron County to Caseville. Promotion for the railroad began in 1879, but was going poorly until

September 1881. As soon as it was learned that the land was cleared by the fire, the shrewd promoters were quick to realize that the land was going to be productive for farming. They announced plans to build the road on September 24, 1881, and named Caseville as its terminal. Grading started the next month, and the first train traveled the track to Caseville on October 3, 1883.

There were other, more subtle changes, which altered the way of life for the pioneer settlers. The wild game that once abounded was gone, and hunting, which provided a major part of the diet for the people, was forgotten. Gone were the herds of elk and deer. The black bear disappeared, as did the beaver.

The women of the area were given their first taste of store bought clothes in the relief packages that came to them. Mrs. Matilda Irwin, who survived the fire as a small girl near Deckerville, remembered those new dresses. She recalled that women usually made their clothes by duplicating their old dresses on new cloth. Consequently, they knew nothing of new styles and were astounded when they received clothes with ruffles and lace.

It was a strange place to live in the years following the fire. Charles Swayze of North Branch tells how his grandfather, Russell Goff, found the 80 acres of land he settled south of North Branch in 1883:

> It was a lonely place, no trees, no birds. The only wild animals were wild rabbits. Many huge fire-blackened stumps were left from the lumbering that had been done before the fire. Also many tall dead stumps that were pitchey and dry were often struck by lightening, . . .
>
> The brush was thick; lots of black raspberry, red raspberry and blackberry bushes had sprung up. Girls

This is one of the few pioneer log cabins still standing in the Thumb of Michigan. It is on the Lawler farm in Carsonville. Courtesy Mary Lawler.

picked blackberries and walked to Beechville [the old name for North Branch] and sold the berries for seven cents a quart.

The people ate lots of rabbit meat, hunting them with a club, a bag of some kind, plus a ferret. People had to drive miles to get apples.

There was one other significant change in the Thumb, the departure of the Indians.

The late Mrs. R. C. Allen, a longtime resident of the Harbor Beach area, wrote of the Indians in a settlement in western Sand Beach Township:

> There was quite a group of Indians making baskets and bringing them in to sell. . . . They were there when I was seven so they must have been there before the fire of '81. But after the fire, every Indian in the county came down and got into their canoes. I saw a line of canoes going down, all their tents and everything piled in the canoes and their equipment, paddling down the lake to Walpole Island in the St. Clair River.

Bibliography

Articles

Bismark, Jerome R. "The Burning Thumb." Student paper, St. Paul's Seminary, Saginaw, Mich., 1966.

Callery, Douglas. "The Fire of 1881." Student paper, History Department, Eastern Michigan University, May 7, 1964.

Maschke, Ruby. "Ocean of Fire: Port Hope Michigan," in "A Wind Gone Down: Fire and Ice" pp. 34–46. Michigan History Division, Michigan Department of State, 1978.

Michigan History Division. "A Wind Gone Down: Fire and Ice." Lansing, Mich.: Michigan History Division, Michigan Department of State, 1978.

Parks, Roderick. "The Thumb Fire of 1881."

Wilson, Bethany. "It's an Ill Wind." *Michigan Alumnus* Dec. 8, 1956, pp. 56–62.

Books

Allen, Ida Tucker. *Remembrances of Mrs. R. C. Allen.* Unpublished manuscript, 1949. This can be found in the Bentley Historical Library in Ann Arbor and the Harbor Beach Public Library.

American Red Cross Chapter No. 1. *Clara Barton and Dansville.* Dansville, N.Y.: Fa. A. Owen Publishing Co., 1966.

Bad Axe Golden Jubilee, 1885–1935. Bad Axe, Mich.: Tribune, 1935.

Bailey, William O. *Signal Service Notes: Report of the Michigan Forest Fire of 1881.* Washington, D.C.: Secretary of War, l882.

Barton, Clara. *The Red Cross in Peace and War.* American Historical Press, 1903. This can be found in the library of the St. Clair County Branch of American Red Cross in Port Huron, Michigan.

Cole, J. F., and W. J., eds. *A History and Record of the Family of John M. and Susan Seder Cole.* Sanilac County, Michigan: privately published, 1939.

Cornish, John J. *Into the Latter-Day Light.* Independence, Mo.: Herald House, 1958.

Evans, David. *Report to the Secretary of the Treasury.* Washington, D.C., October 18, 1881.

Fire Relief Commission. *Report.* Lansing, Mich., 1882.

Gallery, Douglas. *Tuscola, Huron, Sanilac, St. Clair and Lapeer Counties.* 1964. This can be found in the Harbor Beach Public Library.

Gastive, Gustove R. *The History of the American National Red Cross.* Vol. II, *The Barton Influence, 1866–1905.* Washington, D.C.: The American Red Cross, 1950.

Hey, Chet. *Huron County History.* Published by Chet Hey, 1932. This can be found in the Bad Axe Public Library.

Hey, Chet, and Eckstein, Norman. *Huron County Centennial History, 1859–1959.* Harbor Beach, Michigan: The Harbor Beach Times, 1959.

Holbrook, Stewart H. *Burning an Empire: The Story of American Forest Fires.* New York, N.Y.: Macmillan Co., 1952.

Landon, Fred. *Lake Huron.* New York, N.Y.: Bobbs-Merrill Co., 1944.

Logbook, Point Aux Barques Michigan Light Station #2, District No. 10. Sept. 4–9, 1881. This can be found in the Harbor Beach Public Library.

Logbook, Revenue Steamer "Perry." Sept. 4–8, 1881. This can be found in the Harbor Beach Public Library.

Michigan Department of Conservation. *Forest Fires and Forest Fire Control in Michigan.* Lansing, Michigan: Michigan Department of State, 1950.

Moore, Charles. *History of Michigan.* Chicago, Ill.: Lewis Publishing Co., 1915.

Portrait and Biographical Album of Huron County, 1884. Chicago, Ill.: Chapman Brothers, 1884.

Portrait and Biographical Album of Sanilac County. Chicago, Ill.: Chapman Brothers, 1884.

Port Huron Executive Committee of Relief. *Report.* Port Huron, Michigan: Saturday Tribune, 1882.

Raymond, Oliver. *Shingle Shavers and Berry Pickers.* Port Sanilac, Michigan: privately published, 1976.

Schultz, Gerald. *Walls of Flame.* Published by the author, 1968. This can be found in the public libraries of Bad Axe and Harbor Beach, Michigan.

Talbert, George. *Diary.* 1877.

Thompson, Charles E. *Road of Relief of Victims of Fire of 1881 in and around Bad Axe, Michigan.* This can be found in the Burton Historical Collection, Detroit Public Library.

Thornton, Beatrice Cole, and James F., eds. *A History and Record of the Family of John M. and Susan Seder Cole,* Sanilac County, Michigan: privately published, 1956.

Letters

Gifford, Elizabeth. Letter to her mother, Sept. 8, 1881.

Hubbell, Julian B. Letters. Oct. 6, Oct. 7, Oct. 9, Oct. 11, 1881.

Shepherd, John V. Letter to M. H. Allardt, Sept. 7, 1881.

Newspapers and magazines

Detroit Evening News
Detroit Free Press
Detroit Post and Tribune
Frank Leslie's Illustrated Newspaper
Harbor Beach Times
Harper's Weekly
Huron County Tribune
Huron Times
Lapeer Clarion
Mayville Monitor
Minden Post
Port Huron Daily Times
Port Huron Times Herald
Red Cross Courier
Tuscola County Advertiser

Afterword

The *Port Huron Times Herald* has written more about the Fire of 1881 than any other newspaper. These newspaper accounts appeared not only immediately after the fire but over the past century as well. One hundred years later the topic is still newsworthy.

The *Huron Times* issue published Sept. 8, 1881 can be found in the City Hall in Harbor Beach. This issue provides a very detailed account of the fire in the Harbor Beach area, the biggest local event in Huron County. The account of the fire does not appear on the front page, exclusively devoted to national news. Buried in the middle of the paper was a long account of the fire including lists of people who had been killed by it.

The Logbook of the Point Aux Barques Michigan Light Station is the most authentic record of what occurred at the light house and the surrounding area during the six-day period of the fire. On Sept. 8, 1881, the exhausted crew at the light house was confronted by a second emergency: a ship was wrecked in Lake Huron, a few miles from the light house.

The National Archives and Records Service in Washington, D.C. provided a copy of the Logbook from the Revenue Steamer "Perry" from Sept. 4–18, 1881, and the report made by Capt. David Evans to the Secretary of the Treasury Oct. 18, 1881. He detailed the "Perry's" activities in delivering cargo from Detroit to the sufferers in the Thumb area. On Sept. 14, ten days after the disaster, the crew of the "Perry" unloaded "450 bales and packages for the sufferers" at Port Sanilac, Forester and Sand Beach. The "Perry" covered a total of 772 statute miles bringing relief to the Thumb area. The ship was in Lake Erie the day of the fire. The record indicates that even after a week had passed, the smoke over Lake Huron caused severe visibility and navigation problems.

The *Remembrances of Mrs. R. C. Allen* is quoted in a number of places in this book. Mrs. Allen had very vivid memories of the Fire of 1881 and of its aftermath. In 1949, while in the Harbor Beach Hospital, Mrs. Allen repeated her life's story to Lena Attwater of Port Huron who took down the story in short hand. When the author's search for information became known in the Harbor Beach area, Mrs. Allen's excellent account was brought to light. Two of Mrs. Allen's daughters reside in Harbor Beach, Monica Jirasek and Ruth Foe. Allen Jirasek, son of Monica, died while this book was being written. Few people in this area knew more history than Allen Jirasek. He deserves special mention for his advice and assistance.